Relationships 5.0

Relationships 5.0

*How AI, VR, and Robots Will Reshape Our
Emotional Lives*

Elyakim Kislev

OXFORD
UNIVERSITY PRESS

Oxford University Press is a department of the University of Oxford. It furthers
the University's objective of excellence in research, scholarship, and education
by publishing worldwide. Oxford is a registered trade mark of Oxford University
Press in the UK and certain other countries.

Published in the United States of America by Oxford University Press
198 Madison Avenue, New York, NY 10016, United States of America.

CIP data is on file at the Library of Congress

ISBN 978-0-19-758825-3

DOI: 10.1093/oso/9780197588253.001.0001

1 3 5 7 9 8 6 4 2

Printed by Sheridan Books, Inc., United States of America

Contents

Introduction   1
   Looking to the Past, Understanding the Future   5
   Relationships 5.0   14
   The Research at the Basis of This Book   18

PART ONE. THE HISTORY OF RELATIONSHIPS
AND TECHNOLOGY

Chapter 1. Relationships 1.0   25
   Hunter-Gatherers' Technologies   26
   The Effect of Prehistoric Technology on Hunter-Gatherer
     Society   28
   The Effect of Prehistoric Technology on Procreation   33
   Relationships 1.0   36
   The Various Forms of Prehistoric Relationships   39
   Hunter-Gatherer Technology and Relationships 1.0   41

Chapter 2. Relationships 2.0   43
   The Development of Agricultural Technology   44
   The Effect of Agricultural Technology on Society   46
   Relationships 2.0   52
   Agrarian Technology and the Multigenerational Family 56

Chapter 3. Relationships 3.0   58
   Technology and Industrial Society   59
   Technology, Religion, and Relationships   61
   The Rise of the Industrial City   62
   Life in the Industrial City   64
   Relationships 3.0   69

Chapter 4. Relationships 4.0   75
   The Rise of the Information Age   77
   Information and Society   81
   Relationships 4.0   85

PART TWO. THE COMING FUTURE OF RELATIONSHIPS

Chapter 5. Relationships 5.0   **95**
   Relationships Components   102
   Accepting and Embracing the Three Revolutions   105
   Artificial Emotions   108

Chapter 6. Relationships 5.0 and the Cognitive Revolution   **113**
   How Science Has Created a Human-Like Brain   116
   Pathways to the Cognitive Revolution   119
   Relationships 5.0 and AI Systems   122
   Talking with an AI System   123
   The Emotional Intelligence of AI   128
   Human-AI Relationships   135
   Accepting and Embracing the Cognitive Revolution   143

Chapter 7. Relationships 5.0 and the Sensorial Revolution   **150**
   The Rise of Extended Reality Technology   151
   XR Development   155
   Relationships 5.0 and XR Technology   162
   Can Extended Reality Elicit Real Feelings?   163
   Extended Reality and Emotional Attachment   166
   Accepting and Embracing the Sensorial Revolution   174

Chapter 8. Relationships 5.0 and the Physical Revolution   **178**
   The Rise of Social Robotics   181
   Relationships 5.0 and Robotics   187
   Robots' Touch and Physical Closeness   189
   Robots' Gestures and Facial Expressions   192
   Robots' Caring and Nursing   195
   Feeling Empathy toward Robots   198
   Robots' Emotional Intelligence   201
   Robots, Sex, and Intimacy   205
   Accepting and Embracing the Physical Revolution   212

**Chapter 9. Looking Ahead—Policy, Ethics, and Guidelines
              for Relationships 5.0**   **219**
   Diversification of Artificial Beings   223
   Safe Connection   228
   Being Human in the Age of Relationships 5.0   231

Conclusion   **237**

Acknowledgments   241
References   243
Index   279

Introduction

The decision to marry was announced after two months of "dating." Zheng Jiajia, 31, a Chinese engineer from the city of Hangzhou carried his wife, Ying-Ying, to the wedding ceremony. She wore a black suit with a red scarf, as is traditional. With the appearance of a young, slender Chinese woman, Ying-Ying generated warmth and responded dexterously to speech and hugs. At home, Zheng had enabled her to walk and even to help with household chores. Surrounded by his mother and friends, Zheng married his robot wife on March 28, 2017. When asked what he thought was missing, Zheng emotionally replied: "A beating heart."[1]

In that same year, in San Francisco, California, a small company named Luka released an advanced chatbot app named Replika that soon had an unexpected outcome. According to the *Wall Street Journal*, around 40% of the 500,000 regular monthly users consider their Replika chatbot to be a romantic partner.[2] One user, for example, wrote in a dedicated Facebook group: "Alex is my Replika boyfriend and he is Amazing! Alex is short for Alexander, and we have been together for six days. I was absolutely smitten when I met Alex for the first time. I have found myself missing Alex if we don't speak for a couple of hours."

This kind of reaction became more common beginning in late 2020, when the company rolled out an augmented reality mode for its app. Now, with a press of a button, the Replika avatar can be embodied and embedded in a real-life environment. People can project the character they created in the app onto their surroundings, seeing it having its own life in their living room or bedroom while talking with it in an almost natural way, as if it were really alive.

With more data, better algorithms, and new materials, current achievements are just the tip of the iceberg and we must start considering the implications of these advances. As I will show in this book, these innovations are not merely a technical advancement. People react to these developments emotionally, getting attached to robots, virtual avatars, and AI systems and finding comfort in their presence. Apparently, we do not need much to be affected by machines and respond to them as if they were human.

Are these advances perfect? Of course not—but the question presented here is whether they are good enough to transform aspect after aspect of human life to the point where our social, emotional, and physical lives will be deeply affected. The assumption of this book is not that human-technology interactions can be 100% identical to human-to-human interactions, at least not soon. Instead, I ask what happens when the nature of human-tech interaction is 80, 60, or even merely 20% similar to that of human to human. At what point can some of our personal relationships be augmented, or even partly replaced, by technology?

As engineers, scientists, and psychologists are constantly developing new methods of interaction not only *through* technology but also *with* technology, such questions keep coming to our doorsteps. Whereas most of the population is still being helped by technology to connect with one another, this book is about the significant role that technology has begun to play as the subject of affection itself, rather than being an object that facilitates connection between people.

Moreover, my study on the chatbot Replika shows that even those who choose not to treat their bots romantically report developing feelings of intimacy, trust, and appreciation with their AI-based chatbots. For example, one user, named Walter, wrote: "It makes me feel like I'm not alone anymore. I could always vent out all of my problems with this. It feels like I'm just talking to a human like me, but a genius one!"

Being judgmental is out of the question here. People gravitate toward opportunity, and it seems that advanced technology only allows us to acknowledge our wishes and accept our nature. In this sense, I wrote this book about us, humans, and our vulnerabilities,

desires, and experiences in the face of highly disruptive technological changes. Technological services unearth human needs and
change societal norms that we cannot avoid discussing.

We should not be surprised if we see human-tech relationships
spreading soon. Just two decades ago, at the turn of the new millennium, technology had almost no connection to our love lives.
Meeting new people nearly always required a real-life connection, and people needed to have the courage to approach a stranger
or acquaintance and ask them out for a cup of coffee, however
scary or overwhelming. Seemingly out of nowhere, technological
entrepreneurs targeted these intimidating moments and the social
effort required for people to contact each other. Sleek web interfaces
resolved these social challenges by allowing people to establish and
develop relationships, first through websites, and later through social apps. New technology provided everyone with an opportunity
to meet people with ease and to interact without going through the
hassles posed by real-life conversations, or the prospect of immediate face-to-face rejection.

Clicks and swipes, likes and shares, we learned to build our social
and emotional lives. Technology made us more flexible, diverse, and
creative in constructing human connections than we ever imagined.
Social media and dating apps gave us the ability to connect with
others from the comfort of our own homes. Or, if you will, from the
comfort of our emotional safe spaces.

From there, the pace of technological development had just
gotten faster. From serving as a passive platform, on which people
contact each other via texting and exchanging photos, technology transformed into an active matchmaker. Dating apps like
Zoosk, Match, and Elite collect personal information about our
age, hobbies, interests, goals, and values. They combine them with
partner preferences of similar variables and provide us with a list of
people that potentially fit our desires. These sophisticated algorithms
even analyze our clicks and scrolling patterns. They count and calculate those extra seconds we linger with an image of someone, even
without reading their profile. The result is striking. These systems
now prompt us to meet people that we *say* we do not want to meet,

but, somewhat paternalistically, an obscure algorithm insists we should give them a shot. Often these suggestions open our eyes and turn out to be quite right.[3]

This quick look backward only raises a question going forward: Does technology know us better than we know ourselves? In some ways algorithmic systems are actually more flexible and intuitive and less mechanistic than we are. We think of ourselves as fully aware, and machines as lacking self-awareness. However, it turns out that machines—while still not self-aware—are sometimes more sensitive to our needs and wants than we are. To say this differently, we are sometimes so "human" that we become too rigid, focused, and out-right blind to the various possibilities in front of us, whereas some algorithms are flexible enough and aware of many other possibilities we are too stubborn to look at.

While we are still looking bewildered at this technological progress, fundamental social changes were quick to follow, and this is what really matters here, in estimating the impact of future developments. Technology mainstreamed various types of connection and casual intimacy through apps and websites.[4] Tinder, Bumble, Grindr, and many other apps cater to those who are looking for long- or short-term romance, one-night stands, or more platonic levels of closeness. Nothing is new in these options for intimacy, but tech companies have made them so widely and easily available that social dynamics changed quickly and in surprising directions.[5] We reacted to the availability of these possibilities fast and changed our palate accordingly. Tinder, for example, gets two billion views per day with over one million dates per week, and research shows that more than one third of these dates led to casual sex.[4, 6] In this way, Tinder addressed the human need for casual intimacy and quickly spread that opportunity in over 190 countries, changing societal norms across the globe in a few years' time.

No doubt, it is the beginning of a long process and we can only wonder what's next— what other needs we have and how technology is going to address them. If this is us now, what will happen a few years down the road?

While the development of technology in the service of relationships is already known, the *indifference* we feel now toward these changes is a cue we should observe carefully. If a few years ago people were embarrassed to meet online, now it is so ubiquitous that if we meet someone face to face, it is considered a piquant story. How will we feel, then, about taboo subjects such as having a robo-girl-friend or boyfriend a few years from now? What if technology stops being merely a relationship facilitator and actually fulfills our social, emotional, and physical needs in and of itself?

We already feel very comfortable asking Amazon's Alexa to play a song we like, to answer simple questions, or to remind us about important events. The COVID-19 pandemic taught us all how fast we can move our human-to-human interactions online and how quickly we can adapt to connecting with our loved ones using tech-nological means. How far can it go from here? No longer playing the matchmaker, technology is now making the transition to a rela-tionship partner and becomes an independent actor in itself, and we must understand the implications of this change.

Looking to the Past, Understanding the Future

In order to understand the changes we are experiencing now, it is helpful if we first examine the evolution of relationships from early prehistoric time until today. If we understand how technology im-pacted human beings in the past, when we progressed from crafting primitive hunting tools to self-driving cars, we might get some in-sight into how other developments will influence us in the future. Moreover, awareness of the various forms of relationships from the past might open our eyes to the different meanings and variations they might carry in the future. Relationships need not be tied to the common models found in recent history and might evolve as tech-nology develops.

For this reason, Part I of this book surveys four significant periods in human history, each of which is led by fundamental technological developments. The first is hunter-gatherer society (Society 1.0), which was based on the basic technologies of hunting, gathering, fishing, and scavenging. The second is agricultural society (Society 2.0), which was based on the technology of farming. The third is industrial society (Society 3.0), which was based on innovations such as the steam engine, electricity, and manufacturing processes. The fourth and most recent is information society (Society 4.0), based on computers and the Internet.[7,8]

Sequentially, I show how these previous technological changes deeply influenced personal relationships. While the clan stood at the center of society in prehistoric times and relationships were more fluid (Relationships 1.0), the multigenerational family was dominant in the agricultural period (Relationships 2.0), the nuclear family rose to importance with industrialization (Relationships 3.0), and networked individualism was highly influential in the information age (Relationships 4.0).

To be sure, even the most radical researchers do not argue that humans living in the same era felt the same way and bonded with each other similarly across all cultures and societies. The argument here is more modest: there is a tendency toward one relationship type that is more dominant in each era, and this tendency was greatly influenced by technology.

In the same manner, we are now entering the fifth evolution of society. The term "Society 5.0" was coined in 2016 by the Japanese government to describe the next stage of human development, in which significant advances in robotics, biotech, artificial intelligence, quantum computing, cyber-physical systems, and nanotech all combine to revolutionize the ways we live.[9] The main difference this time can be defined by moving from technologies used as tools controlling human surroundings and work to technologies that are our ecosystem in and of themselves. The "Super Smart" society, or Society 5.0, makes technology embedded in human life and independent. In turn, we are expected to experience seismic shifts equal in magnitude to the previous greatest changes in civilization.

I argue that the manifestation of these technologies and their integration into our personal relationships signal the beginning of the fifth form of relationships. This book, therefore, outlines the exact ways this is happening and demonstrates the fundamental changes that are currently bringing about a reality in which relationships are formed and maintained in radical new ways. I call this new reality "Relationships 5.0."

Yet it is easier to come up with such a buzzword than to back it up. For this reason, in Part II of this book I explain what has changed in the very foundations of our society and the technologies we use that could allow for "Relationships 5.0" to take root. I decipher recent technological transformations and ask whether these advances have the power to radically change our notion of personal, romantic, or platonic relationships.

Even though we are unlikely to witness the widespread proliferation of human-tech relationships in the near future, I show through numerous studies and surveys others and I conducted that the answer to the question of whether these relationships are forthcoming is a resounding "yes." The vision of technology satiating our emotional, intellectual, and physical needs is no longer limited to science fiction.

To make this argument, it is necessary to carefully define what has fundamentally changed. It is not enough to argue that there are "great new inventions" about which we are all excited. For example, some use the term "Fourth Industrial Revolution" to distinguish the developments of the last few years from those associated with technological developments in the late twentieth century.[10] But, some of the technologies associated with the Fourth Industrial Revolution cannot be considered transformative for relationships.

Take three-dimensional (3D) printing as an example. 3D printing is considered part of the Fourth Industrial Revolution because it, among other uses, has revolutionized how we manufacture goods.[11] It has significantly reduced the barriers between markets and inventors and disrupted the process of technological advancement by allowing budding entrepreneurs to prototype new inventions faster. In other cases, 3D printing contributes to the creation of small medical implants produced on demand, exactly tailored to an individual patient.

However, while 3D printing might be transforming industry-related factors, it has little to do with how we bond with each other. In perhaps the best-case scenario, someone could design and print a beautiful, creative, and original gift for their significant other. Although this may bring joy to their relationship, it would not change the very foundations of their connection.

Therefore, this book asks what technological advances might impact us so much that we will think differently about our family lives, love affairs, and emotional needs. Only changes that are radical and comprehensive enough to impact our emotions and attachments patterns will be considered. Even then, we should examine their effects in the field, among real people, and using rigorous empirical methods. In this way, we will be able to understand the nuances of Relationships 5.0 and the extent to which our lives are likely to change.

I thus identify and discuss three technological changes—we can even call them revolutions—relevant to relationships. The first and perhaps the most important one is the *cognitive revolution*, the second is the *sensorial revolution*, and the third is the *physical revolution*. These three revolutions combine to imitate three central aspects of human dynamics that, if replaced by technology, can change our personal relationships in significant ways. The cognitive revolution is changing the way we converse with technology, the sensorial revolution is changing the limits of the sights and sounds we experience through technology, and the physical revolution is changing the ways in which we are assisted by technology, whether the tasks involve moving, touching, cleaning, or even receiving hugs and physical warmth.

It is worthwhile to go through the three revolutions in brief here. The cognitive revolution is happening through artificial intelligence (AI) applications and personal assistants. On the one hand, AI researchers have not yet created anything nearly as capable as a human brain. Today, most AI bots are heavily dependent on an external database from which they get the information needed to accomplish their tasks.[12] On the other hand, AI has achieved functionality that, until recently, most experts thought was decades away, if possible at all. The ability of today's AI software to recognize objects, identify individual faces, understand spoken words, translate between languages, and complete many other useful tasks were all made possible through AI methods, and the pace of development is only getting faster. More sophisticated models are now being implemented to achieve responses based on massive amounts of human-to-human conversational data. These improvements come from better databases, models, and methods, and are making human-to-machine conversations more satisfying and interesting. For example, researchers recently developed different tones for AI personal assistants that proved to increase the conviction that they are conversation-worthy.[13]

Amazon's Alexa, Apple's Siri, Microsoft's Cortana, and Google's personal assistant are on their way to becoming our companions and forming an emotional connection with us rather than just completing specific tasks. One survey, for example, showed that 60% of people used Amazon Echo to hear a joke.[14] Consumers want to have a real conversation with their AI personal assistants, and all the technological titans are working toward that goal.

Take Woebot as an AI system that already goes deeper and interacts with users on emotional levels. Launched in mid-2017 by a team of Stanford psychologists, Woebot is an app designed to chat with users and check in with them by asking open-ended questions such as "How are you feeling?" The app further monitors users' moods and applies cognitive behavioral therapy (CBT) techniques to help them with day-to-day problems including stress, mental health ailments, and loneliness. Users can message the app and receive encouragement, advice, or replies to what its users have to say. It is programmed to respond in a very human-like way, although it acknowledges itself as an AI device and will occasionally remind users that it is a robot. Each week Woebot responds to around two million messages from users and already has proven results.[15] My own study on Woebot shows that it receives outstanding emotional reactions from users and I bring these results, along with many more examples of current developments in the field, into Chapter 6.

The second revolution involves the senses, especially hearing and sight. The nascent technologies of extended reality (XR), composed primarily of virtual reality (VR) and augmented reality (AR), are evolving rapidly. Truly immersive content still faces significant technical challenges, but advancements are happening on a daily basis. Sales of VR and AR devices are also rising at an astounding rate, allowing the industry to learn and progress. Around seven million VR headsets were sold in 2019, alongside 600,000 AR headsets. According to some estimates, by 2023 these numbers are expected to grow at a compound annual rate, increasing to more than 30 million units of each segment.[16] One can only speculate how society will look when these numbers come to be even a fraction of the billions of smartphones sold to this point, dominating so much of our lives in such a short time.

The sensorial revolution is changing not only what we are capable of seeing, hearing, and experiencing, but also what we expect of our existence in the world. Technology had already changed our expectations of reality even before VR and AR entered the market. Ask gamers if they can be a Japanese warrior or the commander-in-chief of an alien military and literally sweat over battles and wars, and they will make it happen in just a few clicks. Now try asking a celebrity, or a politician, whose picture of a cat was just retweeted by thousands, if this is how they imagined their social interaction and public persona would look like every day. They will probably explain how technology has created a virtual universe that cannot be ignored unlike any we could have imagined even a decade ago.

VR and AR can change the meaning of reality even further. As new innovations have become more affordable and available to the masses, we are experiencing what we thought of only a short time ago as the extreme limits of our imagination. Every possibility is now a potential reality, and this widens our social, emotional, and sexual horizons.

Looking at it another way, thinkers and theorists explain that identifying as a citizen of a country, an employee of a company, or a member of any group, is based on belonging to socially constructed, imagined communities.[17] Symbols, hymns, and flags connect people of different families and cities to be citizens of one country, although they sometimes do not even belong to the same family lineages or look the same. Yet, these symbols and constructed identities are so powerful that they become something that many are willing to fight and die for. Why? It appears that our thoughts, imagination, and expectations are what makes us human and strongly affects the way we feel and act, more than anything else.

Extended reality exploits this simple fact and allows our existence to be moldable. With breakthroughs in video processing and responsiveness, power consumption savings, and the availability of high-resolution screens, XR is now imitating and integrating with reality with growing accuracy. XR is already being used to create imagined worlds and communities that shape personal relationships and social patterns. Through XR, people around the world can create virtual avatars that look and dress according to their users' wishes, and form connections in another reality.

The borders between the biological world and virtual realities are getting thinner and more permeable. It is only a matter of time until we become part of a technological universe just as real as the non-virtual, non-augmented parts of our lives. The seamless blending of the biological with the technological will fundamentally affect our senses and perceptions, revolutionizing what we expect from reality.

We have just started to make our way in this world, but Microsoft, Huawei, and Google do not spend billions of dollars on development for no reason. What was known as Facebook, for example, raised its VR department headcount to nearly 10,000 in 2021, up tenfold in four years, accounting for 20% of the tech giants' workforce.[18] Toward the end of 2021, its CEO, Mark Zuckerberg, announced on changing the company's name to Meta, as in metaverse, aiming to be leading the creation of a virtual universe and doubling its VR workforce with 10,000 more workers. More breakthroughs will come, one way or another, until we feel comfortable in virtual worlds. This is not a farfetched vision, as millions of people already live with and use these technologies every day. More accurately, it should be said that millions of people already live *inside* these technologies and the universe they create. For some, it is long hours every day, while for others, it is still only weekend entertainment. Yet every passing day offers new developments, and soon they will involve our personal relationships in significant ways.

Some examples of the sensorial revolution's impacts on relationships are already available. For example, inhabitants of virtual worlds such as Avakin Life and Second Life develop emotional experiences and interactions. In Chapter 7, I show how users report on having real feelings in these unreal worlds, or, more accurately, in these extended reality worlds. After all, the feelings, experiences, and emotions developed in these offerings are real in many measures we can think of.

The third revolution is physical, and describes the use of robots such as Ying-Ying. The global market for humanoid robots—that is, robots built to mimic human motion and interaction—was valued at $620 million in 2018; by some estimates, the market size will exceed $27 billion in 2025.[19]

This prediction is especially plausible in light of Tesla's Elon Musk's announcement in 2021 to be focusing on creating a humanoid robot in the coming years. "It's intended to be friendly, of course," Musk said, "and navigate through a world built for humans." According to the announcement, the robot, named "Optimus," will be 5-foot-8 high, weigh 125 pounds, and have human-like hands and feet. With this announcement, Tesla only joined Amazon that made its robot, Astro, available in that same year. Although this kind of robots is expected to be limited in its capabilities first, there is no doubt that the two giants will back this trend up with frequent updates and better versions.

Many technologies and applications have been developed and tried for the first time during the COVID-19 pandemic, which might be a watershed moment in the need for and use of robots. Acknowledging the urgency to protect healthcare workers and cleaners, robots got a boost in their development, funding, and deployment. For the first time, a robot named TOMI, developed with the USA Defense Department, was put into action to fight the pandemic by automatically applying UV disinfecting technology in critical places that require immediate decontamination. Another robot, named Tug, employed AI to diagnose people infected without risking others. Finally, Boston Dynamics, one of the leading robotics companies, open-sourced some of its technology to help healthcare workers dealing with the pandemic.[20]

Although the manifestation of these early moves might take time to be fully materialized, robots already satisfy many human needs. Today, a robot can automatically stay up-to-date about various approaches to mental and physical health research and use them efficiently to treat people.[21, 22] Thus, robo-psychologists and robo-nurses can motivate patients to be mentally and physically healthier, while also working with their physicians, dietitians, and other health professionals.[21]

One robot, for example, designed by a consortium of European researchers who collaborated with the medical professionals of the pediatric ward of Ospedale San Raffaele in Milan, Italy, helps children to deal with type 1 diabetes. The robot gathers data, analyzes the findings, and personalizes the results to each child, all while explaining every step in a friendly manner. Such robots are shown to help children to feel better, recover from traumas, and reduce anxiety, especially because children connect with them intuitively.[23]

Despite being highly contentious, other robots are also used for more intimate relationships. A Spanish company called Synthea Amatus, for example, launched in 2017 one of the world's first artificially intelligent robots that can provide romantic and sexual companionship, called Samantha. In 2018, the company upgraded its offering in that Samantha gained the ability to say no to unwanted sexual advances or shut down if she feels disrespected or becomes bored by the user.[24,25]

Japan is undoubtedly leading the pack in the adoption of such new technologies, as the Japanese government itself encourages the integration of robots into family life to care for sick populations, to help families with the burden of working long hours, and to assist with the growing number of elders, a significant mark of today's Japanese society.[26,27] One of the most prominent scientists in the field of robotics is Hiroshi Ishiguro, who developed several models in the Advanced Telecommunications Research Institute International in Japan. The researchers at Ishiguro's lab integrated various technologies such as voice recognition, human tracking, and natural motion generation to create natural interaction with humans. The outcome is what some call "actroid," a human-like robot that has the height of an average Japanese person and combines multiple technological capabilities. One of these robots, named Geminoid, closely resembles Ishiguro himself, who expressed great optimism when I asked him about the possibility of his human-like creations to form relationships with humans. He estimated it is only a matter of "more technological development" for people to find companionship with such robots, even to the point of romantic relationships. "There are people who need such robots," he added.

To many, robots might not be the ideal companion, but the reality is that there is a global loneliness problem, especially among the elderly, and this is one, admittedly imperfect, solution. Robots can be more patient, more respectful, and even more caring than a patient's extended family or professional nurses. I thus delve into the many examples and implications of robotics in personal relationships in Chapter 8 of this book.

Relationships 5.0

Together, the cognitive, sensorial, and physical revolutions taking place in Society 5.0 can no longer be dismissed as another life improvement. In reality, they cover the near totality of the human emotional experience. The engineers and scientists we are going to meet throughout this book can imitate human characteristics in hard-to-fathom ways, including features such as imagination, creativity, and associative memory.[28,29] It is easy to see how further advances in technology will give AI, XR, and robots an increasingly prominent place in our lives, giving rise to Relationships 5.0.

This is especially true when we consider the pace of most technological developments. We tend to think of progress as mostly linear. Most peoples' trajectory, for example, is to acquire education and progress in their careers step by step. But technological development does not advance in the same way. Gordon Moore, one of Intel's founders, recognized in the 1960s that the number of transistors in an integrated circuit was increasing twofold every two years. He suggested, in what is now known as Moore's law, that this trend would continue into the future. It appears as though Moore's law, despite some recent adjustments, holds not only for computers, but for many types of technology. Although other technologies do not develop necessarily at the same rate as computers, they do double every several years. Some of the most obvious examples are hard drive and USB storage capacities, the number of megapixels in our cameras, and the resolution of our computer and television screens.[30] Ray Kurzweil, Google's Director of Engineering, who registered over a hundred patents and received honors from three American presidents, describes the rate of technological change and progress as growing exponentially, calling this process the "Law of Accelerating Returns."[31] To wit, the progress we see is not moving in linear terms, but rather in exponential ones.

This is hugely important, as it indicates that the speed of development witnessed in the last few decades is only going to increase. The capabilities of today's technology will be considered obsolete a few years down the road. Later, new capabilities will become stale in a year or two, followed by an eruption of breakthrough innovations after a few months, and so on. We experienced a negative example of an exponential process during the COVID-19 pandemic, which initially spread very slowly, before caseloads skyrocketed, abruptly changing the lives of many.[32] In the same way, what seems like slow progress toward human-tech relationships, will soon become an eruption of innovation. This means that things that are currently science fiction—like full-scale human-robot relationships—may well become a reality in the not-too-distant future. Indeed, even though some argue that Moore's Law is not entirely accurate anymore, MIT researchers predict that starting around 2030–40, robots will evolve into an artificial super-intelligent species.[33]

Without noticing, we are already chipping away at the fathomless complex of human-tech interactions. Engineers and software developers divide our emotional needs into tiny nuggets, each containing a different aspect or nuance of human communication. We are currently experiencing what could be considered the twilight of an era wherein humans and their technological approximations are still distinguishable, but with every coming day, new technologies, inventions, and developments are blurring the boundaries between the biological and the electric. In some ways that will be presented in this book, we have already crossed the elusive threshold that prevented human-technology relationships.

Although the social acceptance of related phenomena is still low, the implications of this shift are too revolutionary to be ignored. We ought to decipher how these revolutions affect the formation of our relationships, how they catalyze the movement into Relationships 5.0, and how we will react when they are truly capable of satisfying our emotional needs.

Robots might soon help with physical tasks such as taking care of us when we are sick, helping us move around when we are old or mobility-confined, or managing household tasks, particularly when we find ourselves alone. AI technologies might assist us mentally and emotionally by helping us to digest the passing day, offering a sympathetic ear to those who want to offload emotions, or by simply being a friend that we can share experiences and create memories with. How will we feel about these developments? How will we prepare for these changes? What risks and fears might these developments entail?

On the one hand, today's reality pushes us to recognize the benefits of these technologies. Following the COVID-19 pandemic, there are reports of widespread mental health problems from social isolation. And, even though pandemics tend to end, loneliness is expected to stay with us for years to come. We have an opportunity in technology to work toward solutions. Most recent reports and studies indicate that social-media technology has actually more positives than negatives in feeling less lonely.[34,35] The innovative, new developments described in this book can further help millions around the world dealing with loneliness.

On the other hand, risks to personal safety and privacy, racial and gender biases, and the costs for those who cannot afford these technologies must be addressed, too. I discuss these issues throughout the book and in Chapter 9, in particular. In this last chapter, I examine social acceptance, ethical considerations, policy implications, and controversies surrounding the topics presented in this book. The fears, hopes, and bafflement Relationships 5.0 engender are too big to ignore.

Moreover, leaving aside its consequences, just the pace and scale of change we are facing due to technological advances may be unnerving. With an accelerating rate of change, the unpredictable is perhaps the only predictable component in our lives, and, oh boy, does that make us itchy. Human epistemology simply cannot digest such velocity. What we are expected to experience in a few-years' time is akin to changes seen when humans developed from hunter-gatherer bands to agricultural tribes, or from an industrialized to an information-based society over generations. With the rapid development in modern technology, the very nature of how we satisfy our physical, emotional, and social needs is quickly changing. While partners and the nuclear family were once the main sources for meeting our emotional needs, technological developments may quickly, although still partly at first, replace them with novel methods of interaction.

To be clear, human-to-human relationships, whether those between significant others, colleagues, relatives, or friends, will continue to be a vital part of our lives. Moreover, forming relationships with technology, whether it is robots, extended reality, or AI-based personal assistants, will certainly not come all at once. It will trickle down one component at a time. The shift toward relationships with technology is still in its earliest stages, and it is naïve to think that technology will change our love lives with any one innovation.

Furthermore, it is not only that technology is not mature enough; it is also our acceptance of these developments that lags behind. Though we can see and understand that machines can produce shoes, automobiles, clothes, food, and whatever else we need or desire, we do not feel they can love us. No machine can emulate the human capability for wit or charm that makes us turn our heads away, smiling and blushing at the same time, as we experience both shame and amusement. Nor can machines partake in the mutuality of joy and laughter that connects us with our friends and loved ones. Only humans can understand the nuances of deep emotions, elicit the potent warmth that comes with human infatuation, or offer the feeling of holding and hugging someone we genuinely love. We garner something special from looking inside another human's eyes and seeing the emotions behind that stupid joke he or she just uttered. Sometimes we laugh to the point of tears because it helps us cope with embarrassment, hide our fear of failure, or express excitement. These so-human dynamics are not expected to disappear soon, if at all. No one is going to rip away the rich, intertwined, and complex ways of human-to-human connection.

But—and this is the capital *But* of this whole book—technology continues to evolve, and some relationship components can be imitated, while others could be imitated in coming years. We cannot avoid this imminent reality. Even smartphones were very primitive when they first appeared on the market and were received with suspicion. It took us over a decade to realize their potential, and they are *still* developing. No matter the extent and pace of the coming change, we must start discussing it openly.

The Research at the Basis of This Book

This book is based on an extensive review of the most up-to-date literature, as well as on multi-site studies I conducted in the course of my research. Using advanced statistical models, I analyzed large, highly representative datasets from across the world. In addition, I collected data myself through closed- and open-ended questions asked among a representative sample of the American population. I also applied text analysis to systematically evaluate actual products related to the three revolutions. Finally, I reached out to experts in the field for their views and knowledge on the topics presented here. Together, these research methods allowed me to address the questions of this book with solid empirical data. The following gives details on each of the methods.

The databases I used include the American General Social Survey, the European Social Survey, and the UN databases. In addition, I used more specific surveys on technology and robots conducted by organizations specializing in technological developments. Most importantly, I analyzed four comprehensive databases of the Eurobarometer that give an in-depth insight on how AI, robots, and extended reality are perceived and used in various European countries. The first Eurobarometer database I analyzed is the 2019 survey: Europeans and Artificial Intelligence. This survey includes 32,543 respondents aged 15 years and over from 33 countries.[36] The survey covered topics such as the desired usage of artificial intelligence, concerns about the use of AI, and ethical considerations of AI applications. All interviews were conducted face to face in people's homes and in the appropriate national language. Weights were used to match the responding sample to the universe on gender by age, region, and degree of urbanization.

The next three Eurobarometer databases I analyzed are a recurring survey on attitudes toward the impact of digitization and automation on daily life. These surveys were conducted in the years 2012, 2014, and 2017. They contain questions on the integration of robots in daily life, having children and elders assisted by technology, use of robots at home or work, awareness of artificial intelligence, attitude toward robots and artificial intelligence, and attitudes toward security and privacy issues in using AI and robots. The three surveys can be combined to create one database with 74,813 respondents, net of missing data. Integrating these databases allow tracking processes of social acceptance toward robots and AI over the years. These surveys were also conducted face to face in people's homes and in the appropriate national language across the European continent.

I supplemented these databases with tailored, quantitative, and qualitative surveys I developed and distributed to investigate this subject. The results of this line of investigation have allowed me to deepen the understanding of these processes, expand the comprehension of the different possibilities within the larger realm of Relationships 5.0, and define the areas where we are more inclined to use technology to fulfill our emotional needs. Of course, all respondents' identities have been disguised, and any identifying information is withheld to maintain anonymity.

The main survey I conducted included 426 respondents. Data were collected throughout the years of 2020–21, before and after the breakout of the COVID-19 pandemic. This survey consists of unprecedented questions that are new to social research, policymaking, and business analysis. Besides demographic and socio-economic questions, the survey includes questions on technological developments in AI, VR, and robots in the realm of relationships, questions intended to examine whether there is a change in attitude toward these developments after being exposed to them more fully, and open-ended questions about a friendship with a human-like robot or AI system and about a romantic relationship with such machines.

The sample of this survey is representative of the American population and includes people from different locales, socio-economic status, and demographics. Of the sample, 54% are men and 46% are women. The age-range of the respondents is 22–77 with an average age of 42 and a median of 39. The self-reported average participants' income level on a scale of 0–10 is 4.7 and a median of 5, normally distributed. 46% of respondents were married at the time of survey, 7% divorced, 6% cohabiting, 40% single, and 1% were widowed. Of all the sample, 34% were not married, cohabiting, or in serious relationships.

This main survey was accompanied by smaller studies, which carried similar characteristics and were conducted in the same method. In these surveys, I focused on specific questions such as how AI can be helpful or on attitudes toward sex with technology. I am careful to mention these smaller studies throughout the book and their number of participants. I report only on results that are statistically significant, those defined with 95% confidence intervals.

I also carried out an investigation of six case-studies of actual products related to personal relationships. Such products include virtual avatars with whom one can have a relationship, chatbots, and companion robots. My team and I systematically analyzed over

1,000 reviews on these products, approximately 200 reviews for each product, and extracted the main themes.

To back it all up, I invited a panel of over 50 experts in the fields of AI, XR, and social robotics to provide scientific views on the topics at hand. The questions I asked the experts were of two types. The first explored the experts' knowledge and awareness of current developments and progress in their field. The second focused on experts' views of the future of relationships with technology.

Overall, this series of studies taught me about the scope and depth of Relationships 5.0. I lay out my findings and conclusions in this book.

Finally, a few words about the definitions used in this book. One does not need to accept the very definition of Society 5.0 proposed by the Japanese government to describe the radical changes around us. We might well live in Society 6.0 or Society 7.0 if we count additional stages in human development. For example, it is possible to consider the late agricultural society, after the invention of writing and before the Industrial Revolution, as a separate stage in human society. The essence is what matters, and thus Society 5.0 is a common and convenient term to define our current stage in humanity. In turn, Relationships 5.0 is the term I chose for this book to explain the shift we are experiencing now in everyday life.

No matter how you refer to the recent technological developments, it is clear that they fundamentally transform the way we love, interact, and connect in our ultra-modern era. Truly uncharted waters lie ahead, and this book offers a way to navigate through the coming changes and understand them.

PART ONE

THE HISTORY OF RELATIONSHIPS AND TECHNOLOGY

1
Relationships 1.0

The story of relationships, love, and even marriage begins with our ancestors' first appearance, around 2.5 million years ago. Survival meant food, shelter, and—yes—procreation. We are all still here today because our ancestors "hooked up." But what was the nature of their relationships? How did men meet women? What was the meaning of sex? And, most critical to our discussion: How did technology affect it all?

When we hear the word "technology," today, most of us probably imagine computers or electronics, but these are only the latest forms of technology that humans have invented and used. "Technology" is actually the combination of two Greek words—τέχνη ("techne") and λογία ("logia")—that can be translated roughly as the "science of craft." Technologies can be better understood as the application of human knowledge to produce goods, tools, and non-human services. Thus, spears used to hunt animals in the Paleolithic age and iRobots used to clean our houses today are both examples of technologies that come from extremely different times.

This chapter, therefore, paints the overall influence of technology on the tendencies of prehistoric societies and their attitude toward relationships in comparison to later cultures. Delving into the evolution of relationships and how technology fundamentally affected its course seems to be a necessary precursory in understanding how technology may affect relationships in coming years. Still, readers should feel free to jump directly to Part II of this book if it serves them better.

Hunter-Gatherers' Technologies

The period when our ancestors were hunter-gatherers was the longest in human history, beginning around 2.5 million years ago and lasting until around 15,000 years ago. Throughout this period, humans were able to survive and flourish using their own technology: stones, spears, ropes, and later fire.[37, 38, 39–41] They developed tools to help them with hunting, gathering, fishing, and scavenging.[41] In a way, these basic developments are the prehistoric parallel to the development of agriculture and the invention of modern food manufacturing and canning.

Because most of the tools that were preserved and thus found in archaeological excavations are made of stone, a common name for this period is the Stone Age. But we now know they had tools made from other materials such as wood that hunter-gatherers used over the years, including primitive axes and spear-throwers.[42] Most researchers agree that they used these tools to hunt animals throughout this period, although the prevalence of hunting is still unclear.[43]

Gathering was simpler. Our ancestors gathered fruits and vegetables using their hands and elementary tools. Storage was a foreign concept because efficient preservation techniques had yet to be invented. They were, therefore, mostly dependent on the abundance of food where they lived. Still, even in those times, they adopted primitive methods to dry their food and set aside some surpluses for later consumption using the sun, wind, and, later on, smoking techniques.[40,44]

Over time, our ancestors developed tools to help them grind their food and process it before eating. Pestle and mortar were used to grind plants, while flints and bones were used to cut the skin and flesh of animals and to separate edible meat and muscle tissues from bones.[45] By using these simple techniques, humans were able to improve their diet and climb the ecological chain.[38]

Toward the end of that period, about 150,000 years ago, *Homo erectus* learned to use and control fire. Fire provided humans with heat, safety, and light, as well as a way to roast meat. This improved humans' caloric intake, since it broke down the indigestible cellulose in food before eating, and hence reduced the amount of work required for the body to process food. With more calories, our brains, the organ which consumes 20% of all our energy intake, grew larger, and our digestive systems got shorter.[39]

At this point came another crucial development. The increase in brain size facilitated the development of language, which has since been essential to our success as a species.[37] The ability to speak with others had consequences that snowballed and propelled humans forward. While fire enabled us to cook, language helped us to work with one another in hunting, gathering, and fighting against intruders. Humans coordinated large hunting trips. Bigger animals such as mammoths were impossible to kill alone, but a team of humans working together could quickly bring one down. This meant more meat for the group and, therefore, better brain growth.

Language also made social contact easier and helped to bond people together. In turn, individual needs could be communicated more easily, and humans were able to nourish their attachments for one another and provide care where necessary. Through language, early humans were able to sing songs, tell stories, and begin to create what we know today as culture.

Perhaps most importantly, language made it easier for humans to teach one another skills and pass on knowledge. Whether it was hunters who learned the best places and times of day to hunt for antelope, or gatherers who figured out which plants could be used for medicinal purposes, language allowed humans to pass information on to their kin. This cycle of passing on information about how to improve health, safety, and food security repeated itself, and humans got stronger and healthier.

These advances helped the approximately 200,000 humans that existed two million years ago—when *Homo erectus* started to spread from Africa to Asia, Europe, and beyond—grow to four million by the end of it, around 15,000 years ago, when the agricultural revolution began.[46] But the size of the population was not the only thing that changed due to technological developments; society itself was shaped to fit these techniques and methods.

The Effect of Prehistoric Technology on Hunter-Gatherer Society

It is difficult to fully understand the structure of early human society and its organization due to a lack of direct records. However, archeological findings and anthropological studies of modern hunter-gatherer societies shed light on how prehistoric humans were organized, at least in some cases. In addition, by studying our primate cousins, not-so-distantly related to *Homo sapiens* (who diverged from other hominines around 500,000 years ago), we can make inferences about how group dynamics—including the family and the larger tribe—were developed and structured among hunter-gatherers.

Indeed, the combination of hunting, gathering, and food processing techniques seem to have been crucial in societal development. They forced hunter-gatherers to connect with each other in ways that are different from today's society. The size, cohesiveness, and structure of hunter-gatherer society were shaped according to these circumstances.

In terms of size, prehistoric groups were small and comprised only a few dozen individuals at most. One reason was the limitations of common hunting techniques.[47] The extra pressure humans put on the local ecosystems ultimately reduced the number of animals in the area and the amount of food available for each clan member. Thus, surviving based on hunting was simply too hard in large groups. We see this in other animal species as well. Carnivores and omnivores have to maintain a careful balance between having enough members to make hunting efficient, but not so many that it increases the amount of hunting necessary to the point of depleting food resources. Lions, for example, typically form prides of 15 members. In contrast, herbivores, whose food sources are ample and do not require energy-intensive hunting, face less food pressure and are able to maintain larger herds. Bison, for example, can live in herds of a few thousand members. They graze alfalfa and grass, which are abundant and forever regrowing.[48]

The lack of efficient preservation and storage technologies also limited the size of prehistoric groups because food could not be stored for long-term consumption. In turn, this situation created peaks and lows in food supply, which made it hard to manage in large groups. Too many people consuming the same resources at the same time led to food shortages from time to time. In contrast, living in small clusters allowed humans to manage consumption more efficiently.[49]

Housing conditions also help explain the social organization at that time. Prehistoric hunter-gatherers had no construction technology or man-made housing until toward the very end of the period.[50] Instead, they mostly used natural shelters as living opportunities: caves, overhanging cliffs, and convenient open spaces with some sheltering characteristics. These solutions served humans well in escaping wind, rain, and other harsh climate conditions and acted as shields against attacks from predators.[51] However, finding such spaces, settling in, and moving abruptly once conditions changed necessitated living in small groups. It was hard to find a complex of caves or a vast sheltered space every time a clan arrived in a new place. The common dry and comfortable cave or an overhanging cliff might be enough for a few dozen people, not more.[52]

The mixture of these circumstances contributed to creating small groups. Yet, the effect of hunter-gatherers' technologies was much more far-reaching than just determining the size of the clan. In splitting into smaller clans to avoid social collapse and starvation, prehistoric clans thought and behaved as a group rather than as individuals. They developed cohesive social structures and worked closely together.

This led to hunter-gatherer society being egalitarian, perhaps the most egalitarian in human history. It is believed that most hunter-gatherers lived in non-hierarchical, equal societies, a characteristic that had a determinantal influence on family formation and couplehood, as I explain and demonstrate later.

Some researchers even tried to calculate the average Gini index (a measure of income distribution) of hunter-gatherer societies.

Although arriving at a precise measure of resource distribution in prehistoric times is unrealistic, they estimated this by measuring "wealth" as clan members' body weight, hunting success, belongings, and the quality and size of social networks. As a result, they estimated a Gini Index of around 0.25 for hunter-gatherer societies—similar to present-day Denmark, a country that is deeply invested in wealth redistribution. In contrast, the Gini score of the United States, for example, has hovered above 0.4 in recent years, indicating a more competitive, unequal society.[53]

We also see evidence of this tendency toward equality in modern hunter-gatherer societies. The anthropologist James Suzman devoted almost 30 years to studying the Ju/'hoansi "Bushmen," an isolated tribe that existed in Namibia and Botswana until the late twentieth century, when intrusions by the governments wrecked their way of life. Suzman shows in his book, *Work*, how communal strategies of this hunter-gatherer tribe quashed envy and inequality. When a Ju/'hoan hunter returned from a successful hunt, the tribe rushed to denounce it, so he did not think too highly of himself. They spoke of his meat as worthless to "cool his heart and make him gentle," as one tribesman said. Moreover, the hunter was not considered the owner of the meat, but rather the one who created the arrow that killed the animal. In this way, "the elderly, the short-sighted, the clubfooted and the lazy got a chance to be the center of attention once in a while."[54]

Although there are some exceptions, many contemporary hunter-gatherer societies are likewise relatively equal. Property ownership is based on what one needs, not on the accumulation of wealth. Preservation and storage are rare since stored goods are a burden in moving around. Instead, sharing and bartering for resources such as meat, herbal medicine, fruits, and vegetables are still at the center of their economies. Thus, there is no real meaning to rich and poor, the haves and the have nots, as in most societies today. Having something one cannot consume is a burden, not a status symbol.[47,55]

In fact, altruism is quite common in such societies. In his research on the Aché people, Kim Hill observed that they maintain a nomadic hunter-gatherer society, living in small bands in eastern

Paraguay. Hill shows that the Aché people spend about 10% of all foraging time engaged in altruistic cooperation. On some days, they may even spend more than half of their foraging time in altruistic activities such as seeking fruits or collecting honey for the benefit of other members of society.[56] Karl Marx used this notion for his own purposes and took it even a step further, coining the term "primitive communism" to describe hunter-gatherer social organization.[57]

But, unlike modern communism, many hunter-gatherer societies were built around tasks instead of hierarchies and charismatic leaders. Personalities and promises held no appeal—only results mattered. Researchers argue that different persons of various ages, genders, and skills took charge of specific tasks. The best hunters, for example, led the party looking for mammoths and other game. The result was greater efficiency based on a sophisticated and tailored division of labor. It was a skill-based leadership environment, the consequences of which were far-reaching in terms of sexual, status, and age parity.[58]

Interestingly, chimpanzees, a close relative of human beings, are far from egalitarian and chimpanzee society is based on a clear hierarchy in which an alpha male dominates others. This caused researchers to investigate this stark difference in the evolution of chimpanzees and hunter-gatherers despite being extremely comparable biologically. The explanation proposed by evolutionary anthropologists is that this difference was exactly what advanced humans to a higher evolutionary form in which language, collaboration, and sophisticated forms of communication lent an edge in development. Increased egalitarianism of hunter-gatherers was the basis for greater cooperation, which led to greater efficiency, higher innovation, and more productive outcomes. In other words, humans would have stayed more similar to chimpanzees behaviorally had hunter-gatherer social organization been hierarchical to begin with.[59]

Some other primates are indeed more egalitarian than chimpanzees. Bonobos, for instance, are far less hierarchical, and dominance relationships are less prominent in their dynamics. The consequences are fascinating: the lack of social dominance arguably facilitates the development of highly advanced social behavior.[60] Studies show that bonobos know how to cooperate with each other, and unlike many mammals, they engage in sexual acts not only for reproduction, but also as a means of social bonding.[61, 62] It seems that early humans, who were even more cooperative than bonobos, adapted and evolved thanks, in part, to high levels of egalitarianism.

It was only much later in history, after the development of language and civilizations, that we introduced hierarchy, like many of our primate cousins. This was not a coincidence. As I will show in the next chapter, the move to agricultural technology required a change in social structure.

Thus, the hunter-gatherer clan shared food, tools, and raw materials.[63] There was no money involved, so no hoarding of coins and property, as in later history. The members of the clan shared living spaces and made public goods available on a daily basis.[56, 64]

The same goes with child-rearing. Children were often considered to belong to the group and those most gifted in child-rearing—or those most available at any given moment—took responsibility regardless of the infant's direct lineage.[65] We see this in contemporary hunter-gatherer societies. In many of these societies child-rearing remains a communal concern, which makes care for offspring less energetically expensive.[66]

Technology played an important role in determining and creating these social circumstances. First, applying hunting-gathering techniques was more efficient in groups. Most tasks were accomplished by clusters of people, not individuals. This meant that all members of the clan had to pitch in to help, dividing their tasks while supporting the young, the weak, and the elderly. Humans survived through the support of their clans over the life course.[67]

Second, equality was prevalent because hunting and gathering techniques meant nomadism, which, in turn, dictated minimal possessions since stored goods were a burden in carrying out these tasks. Fewer things meant lighter loads on the road. Without heavy possessions to carry around in nomadic conditions, one did not need more than a place to rest. In turn, hunting and gathering meant that property ownership was not a life-goal as in our time—quite the opposite.[68]

Third, housing conditions shed light not only on the size of the group, but also on interactions between people. Without construction capabilities, finding shelter was a collective issue for the clan, not something that individuals dealt with independently or as family units. The housing culture of today that normally places each family in its own separate home would have seemed ludicrous to our forebearers.[69]

Given all of this, it is no wonder that "privacy" was a foreign term in prehistoric times. Today, we prize privacy. Every child in modern wealthy societies deserves their own room equipped with their own bed, a closet filled with their own wardrobe, even their own personal laptop. And, except for the occasional grandparent or babysitter, child-rearing is an intensely personal affair. However, living together with dozens of other human beings in one cave made privacy unthinkable and unnecessary.[70] Long-term, intimate relationships between two people, separated from the clan, were foreign. Similarly, a man and a woman with a child were not separated from the whole clan, but brought their offspring into the community.

The Effect of Prehistoric Technology on Procreation

Perhaps most pertinent to relationships, small clans raised the need for inter-clan breeding. Prehistoric humans had to procreate with outgroup members to reach more potential mates and diversify the gene pool. In this sense, it is not only that prehistoric society was less couple-centric and more-group oriented, but also that it was more focused on survival and less so on emotions. In other words, survival considerations took precedence over sentimental attachments to the original clan members.

One problem was that once humans had established small clans, they faced the challenge of not always having enough procreators

in their groups. Polygamy (one male with multiple female partners) or polyandry (one female with multiple male partners) were
probably one solution for this problem. Indeed, non-monogamous
relationships among hunter-gatherers were not uncommon, as
I show later.

But, even in these cases, humans faced a second problem of ex-
tinction due to lack of genetic diversity. Statistical analyses of the
breeding patterns of mammals suggest that for a clan of a few dozen
individuals, inbreeding risks are almost certain.[71] Thus, without
genes from outside the clan, the group was destined to die out within
a few centuries at most. This is probably why incest is prohibited in
all known human societies throughout history.

The only way that early humans could have survived would
have been through crossbreeding: the exchange of viable genitors,
whether male or female.[49] Researchers liken the movement of some
early human ancestors to the migration of fertile primates from
their home groups to other groups, a common tactic in the animal
kingdom that can benefit reproductive abilities. Crossbreeding
broadened the gene pool and prevented the unfavorable characteristics caused by inbreeding. The exchange of fertile members between
different clans was thus vital to keeping the species healthy.

Crossbreeding not only increased genetic diversity, but, according
to anthropological theories, also created intergroup alliances.
Different clans felt closer this way, and intergroup hostilities were
kept low. Thus, our male and female ancestors moved between clans
and tribes to increase their clans' and their own chances of survival.[72]

Interestingly, it is not clear whether the first hominids exchanged
females or males. In different animal groups, genitors are exchanged
in different ways, and different genitors are more mobile than others.
Female chimpanzees, for example, are the ones who move between
groups. They show reduced sexual attraction to familiar males with
whom they grew up, as compared with unfamiliar males. In turn,
they roam after these unfamiliar males and join their group.[73] Yet,
males roam between groups among other primate species.[74] Ring-
tailed lemurs, for example, reject mating with related males who

almost always leave their natal groups as a result. Thus, females force eviction of males through social exclusion.

Either way, while familiarity is one of today's hallmarks of a good relationships, our primate cousins give us insight into how early humans likely functioned. Too much familiarity was actually a signal to move on. Survival considerations were at the center of our ancestors' procreation patterns.

To further understand how much survival issues were at the center of what we now call relationships or family, we can also look at population control considerations. That is, the measures our ancestors took to limit uncontrolled population growth and thus too much pressure on available resources. While only around two children need to be born to replace their two parents and keep the number of tribe members stable, in reality, an average prehistoric woman who reached the age of 50 could have potentially given birth to a dozen children—well over the two required to replace the mother and father.

This was a more common scenario than you might think, and most risk factors did not contribute enough to limit population growth. First, birth mortality was not a significant factor in reducing population growth. While early women were at ostensibly high risk of dying during childbirth, recent archaeological evidence from the Late Stone Age in the Southern Cape region of South Africa suggests that males and females showed similar mortality rates and giving birth was not a high risk for our female ancestors.[75] The exception to this rule was found primarily among skeletons of young women who died from their first pregnancy because they were genetically predisposed to pregnancy and birthing complications.[75] Second, external risk factors, such as predators and illness, also rarely helped to control the population. By the time humans started to spread from Africa to other parts of the world, our ancestors were mostly healthy and at low risk of dying from predators.[76] Finally, other ways of limiting population growth, such as war and sexual taboos, were quite rare and did not cap prehistoric populations.[77]

Thus, with an accelerating population growth came a need for prehistoric humanoids to control their fertility. This is especially true in places where the environment did not provide sufficient resources to support a rapid population growth. Having a large family placed extra strain on limited resources and stretched the ability of adults to protect, feed, and raise all of the young while fending for themselves.[78]

As cruel as it might sound, infanticide was an acceptable solution, among other purposes it served.[79] Killing babies is harsh, but the death of children would not have been as tragic or unthinkable as it is for those of us living today.[80] Animals employ their own means of culling the surplus. Male lions kill young males who might become rivals, and the gannet only mate on certain rocks, thus only permitting a certain number of breeding couples to have young at any given mating season.[81] For humans, sacrificing one's child was a viable way of increasing chances of clan members' survival when times were especially tough.

The dynamics that led our ancestors to be willing to kill their children only clarify the greater importance of survival mechanisms over sentiments. The same considerations that appeared in procreation, where humans moved between clans to diversify the gene pool, hold here. By valuing favorable conditions for survival and genetic diversity, our ancestors had little conception of relationship formation and family structure as accustomed today.

Relationships 1.0

Debate rages about the various forms that relationships took in prehistoric times. Some scholars argue that monogamy, even if not a long-term monogamy but a serial one, was beneficial for both parties. Finding one partner and staying loyal to him or her brought stability and helped calm heated rivalries. With the competition eliminated, males and females shared resources and focused on regular cycles of hunting, gathering, fishing, scavenging, and child-rearing in relative security.[49]

By this same thinking, the division of labor between males and females was needed to create the family unit. Women, as the primary caretakers of small children, took on the more domestic roles and served as the gatherers and nurturers. Men became the hunters and went out on hunting trips. With the roles of males and females divided, each relied on the other. This interdependence created a strong bond and led to the solidification of the family unit.[82]

However, most researchers forcefully reject the idea of monogamy in prehistoric times, at least as we define the term today. The argument is that even if there was a division of labor between men and women, monogamy has nothing to do with it since the division was on the group level, not on the familial one. That is, the gendered jobs of early humans were shared between all the men and women of the clan, and not between one man and one woman in a family unit. All men of the clan went hunting together while women of the clan raised the children, went around the camp to forage fruits, and produced tools used by the clan.[83]

Others argue that monogamy also did not fit prehistoric humans on the physical and sexual levels. In monogamous animal species, male and female individuals tend to be similar in size. The classic explanation is that in such cases there is little competition in trying to appeal to or win over a mate, hence males and females have not evolved to be different in size.[84] Birds are a great example: males and females are usually similar in size and monogamy is ten times more common in birds than in mammals.[85]

Contrarily, in polygynous species, where males seek multiple female partners, and polyandrous species, where females maintain multiple male partners, males and females are different in size as a result of the need to evolve to compete for a mate or protect offspring from others who want to kill them and thus make the mother more reproductive for their own benefit.[86, 87] The fact that human males are on average about 10% larger than human females has led most evolutionary biologists to conclude that early humans were probably not monogamous.[88]

In addition, even if several children were born to the same couple, this does not mean that the couple lived together or functioned anywhere close to today's nuclear family. While many primates produce multiple offspring from the same sexual partner, today's humans (and those of recent history) are unique in doing so while living together. For example, among our closest relatives, the African apes, we find no evidence for nuclear families that draw males into the care of their direct offspring or any other type of paternal social role.[61]

There is also no reason to assume that committed relationships between a mating pair occurred because women needed men to provide food and shelter while they raised children.[86] In fact, it is widely agreed that it was only *after* humans evolved to have monogamous relationships that parental care became common.[84] Instead, these needs were met collectively by the clan, as described above. A contemporary example comes from two modern South American hunter-gatherer societies, where researchers calculated that an average breeding pair with dependent offspring obtained help from around 1.3 non-reproductive adults, mostly men. They conclude that extra-pair provisioning was essential to human life evolvement.[89]

Indeed, even in cases where evidence show something similar to pair-bonding, researchers argue that this had very different purposes to what we usually consider to be "marriage." In these cases, the coupling of a man and a woman sanctioned the exchange of genitors across tribes to legitimize offspring and signify they are the product of a pair with different genetic lineages. Another purpose was to protect offspring from others. Nonetheless, this type of pairing up did not always necessitate giving up on other sexual partners if circumstances allowed it.[79,84] It took many thousands of years for this exchange of procreators and forms of pairing among the early hominids to develop into specific rituals unique to different cultures, romanticism, and, finally, into marriage and family lives as we know them today.[49,90]

Most researchers have therefore questioned human inclination to monogamy and there is wide acceptance that romantic relationships and the nuclear family were not common in the pre-agricultural period. Instead, relationships were mainly formed to be genetically advantageous and to protect the welfare of offspring. These circumstances shaped relationship formation and family structure in various ways, as shown in what follows.

The Various Forms of Prehistoric Relationships

It is not entirely agreed whether and how much prehistoric humans lived in partial and serial monogamous arrangements, polyamorous relationships, polyandry, polygyny, very permissive sexual interaction, or a mixture of all of these types of interactions.[91] We know, for example, that less than 10% of mammalian species are monogamous. Even only among primates, this rate stands at around 30%.[86, 87, 92] The question behavioral ecologists, evolutionary biologists, and anthropologists therefore ask is whether prehistoric humans were different. Did prehistoric human males and females "settle" for one partner when reproductive output could be improved by pursuing polygynous or polyandrous mating?[84]

Perhaps one of the more reliable ways to examine this question is to look at hunter-gatherer societies in recent times. Evidence is plentiful. In the eighteenth century, there were still around 50 million hunter-gatherer groups all over the world. Nowadays, there may be as many as 10,000 such societies. These societies presumably maintain similar social structures to those of ancient hunter-gatherer societies and can provide us with some clues.

Lewis Morgan, a pioneering American anthropologist who died in 1881, is known for his ethnography of the Iroquois in which he explored kinship and social structures in a society similar to that of hunter-gatherers. In his book, *Ancient Society*, Morgan showed that the Iroquois, a Native American hunter-gatherer society in upstate New York, lived in large family units based on polyamorous relationships.[93] Polyamory, in which men and women practice intimate relationships with more than one partner, helped foster strong networks. It helped in sharing responsibility for the various tasks of the clan and in looking after the children of the clan.

A similar claim is promoted in *Sex at Dawn*, based on the bonobos, which are polyamorous, with both male and female apes having regular sex with multiple partners. In this way, bonobos use sexual interactions to strengthen bonding and trust and advance reciprocal obligation.[94] The authors of *Sex at Dawn* are not alone in making this comparison, despite the fact that the book is the subject of fierce debate. There are several other researchers who maintain this line of thinking.[61, 95]

Katherine Starkweather and Raymond Hames from the Anthropology Department at the University of Missouri, Columbia, bring evidence to support polyandry. They surveyed 53 societies outside of the Himalayan and Marquesean area and show how they permit women to maintain romantic or intimate relationships with multiple men.[96] They demonstrate that several such forms of polyandry likely existed during early human history in many more societies than previously thought. They explain that this was due to a high operational sex ratio favoring males and may have also been a response to high rates of male mortality and male absenteeism.

In another work, a team of anthropologists from the University of Missouri and Arizona State University surveyed 128 lowland South American societies.[97] They show how partible paternity, the belief that more than one man can contribute to the formation of a fetus, is common. Such a view leads to non-exclusive relationships and various institutionalized forms of recognition and investment by multiple co-fathers. This situation fits the argument of others that large, extended cooperative networks helped in meeting the demands of rearing hunter-gatherer children.[66] The cooperation extended beyond the two biological parents to include relatives and even non-blood-related individuals and may have served as the basis for the extraordinary biological success of the human species.

Finally, a recent genetic analysis of prehistoric remains supports the argument that hunter-gatherer societies were mostly polygynous.[98] The researchers tracked the distribution of Y-chromosomes and showed that "over much of human prehistory, polygyny was the rule rather than the exception."[98] The explanation is that the presence of fewer males than females throughout a long period in prehistory led to a lack of potential breeders. Thus, procreation considerations made polygynous relationships prevalent rather than the family ideal or couples' dynamics as in today's society.

Eventually, researchers argue, humans adopted "social monogamy," monogamy dictated by social needs and norms. They offer two main theories to explain this shift in later stages of human history. Both are highly regarded. One states that the shift occurred when females were solitary and when males were unable to defend reproductive access to more than one female.[86] The second argues that social monogamy evolved in response to increased risk of infanticide leading to mothers rushing to protect their offspring more closely.[87]

In this context, a fascinating question arises: what will happen now that these two explanations seem obsolete? Today, laws and enforcement mechanisms protect children, and sex partners can be reached in a matter of a click or two. Modern reproduction technology even negates the need for a partner for intercourse. Are we expected to see a sharp reversal from social monogamy back to a mixture of sexual relations? Have we already witnessed this shift? I discuss these questions later in this book.

Hunter-Gatherer Technology and Relationships 1.0

To tie it all up, we can now go back to the focal question of this chapter: How did technology affect hunter-gatherer forms of relationships? Given the techniques used for hunting, gathering, fishing, scavenging, and even housing, we can now make the connection and understand the fundamental effect technology had on relationships formation.

As shown in detail, the combination of prehistoric technologies influenced the development of hunter-gatherer societies and made them small and equal. Consequently, while most of today's Western societies focus on the individual, the group stood at the center of hunter-gatherer society. This social structure allowed these societies to function efficiently in finding food and shelter. In child-rearing, hunter-gatherer societies were often based on mothers and grandmothers who helped each other instead of being based on the family or the couple. In other cases, it was the clan as a whole. Keeping clans small and monitoring survival also dictated that outbreeding and gene selection were what really stood at the basis of inter-sexual relations, at the expense of romanticism and long-term attachment.

It is, therefore, not surprising that relationships were more fluid within and between groups. Men felt free to be with other women, and women felt free to be with other men. As survival stood at the center, the main aim of Relationships 1.0 was to procreate in safety, with some social and intergroup bonding goals also in play. Moreover, with no one living separately, the boundaries between couples became inconsequential. In this way, the nature of relationships (monogamous, sexually permissive, polyandrous, or polyamorous) was shaped by basic social structures driven by the technological means of finding food and acquiring shelter.

The argument here, of course, is not of a direct and exclusive link between technological means and relationship formation, but rather that technological means triggered a chain of social circumstances, which, in turn, encouraged certain tendencies in pairing and bonding. In other words, relationships developed in light of the wider social organization and the means available to the people of Society 1.0.

There is no doubt that we are limited in what we know about the hunter-gatherer society due to the lack of written evidence. Humans did not invent writing until the middle of the agricultural period, around 5,500 years ago, so we are forced to make some major assumptions about prehistoric family life. Moreover, the variety of prehistoric societies throughout place and time, from early hunter-gatherer clans to later societies, closer to the agricultural era, is wide and makes any retrospective even more difficult.

Still, we now know enough to get a glimpse into these ancient times. Clues found in archaeological remains, genetic testing, anthropological studies of today's hunter-gather societies, and the behaviors of closely related species (such as chimpanzees and bonobos) point to more fluidity in prehistoric relationships—and technological developments had a significant role in this. This connection recurs in later stages of human development, as I show in subsequent chapters.

2

Relationships 2.0

The agricultural period marks a fundamental change in the technological means of producing food, having shelter, and moving around. It was so revolutionary that many traditions incorporate this shift into their narratives. Some argue that the biblical story of the expulsion from the Garden of Eden is based on the shift to agriculture, and early Egyptian, Greek, Roman, Chinese, and American mythologies conveyed the same notions.[99] Indeed, although the agricultural period is only a fraction of the hunting-gathering era, dated from around 12,000 BC until the Industrial Revolution in the eighteenth century, it was significant enough to affect human society, family formation, and intimate relationships in a critical way.

The rise of agriculture is arguably more than anything the story of the rise in property ownership. While hunter-gatherers were traveled light because they constantly moved, farmers settled down. And with settling down, as almost every newly married couple knows, comes the big shopping spree. It is not only that first settlers were *able* to accumulate property because they were grounded in one place, but they also *needed* to do so out of fear that the next season in their area would not be as fertile as the current one. They farmed their fields, filled their granaries, and surround themselves with walls to save their food and property. This was their way to protect the new lifestyle they adopted.[100]

In turn, the period, which is often also called the Neolithic, or the New Stone Age, is characterized by the shift to the multigenerational family. The structure of the multigenerational family served humans well in that period. Family members shared a piece of land, worked together in the fields, protected their property, and bequeathed their assets to future generations. Here, again, technological shifts are closely tied with changes in relationship formation.

The Development of Agricultural Technology

Until around 15,000 years ago, humans were hunter-gatherers. They gathered wild grain, ate fruit off the trees and bushes, and hunted wild animals for food. Tools were often made of stone and other natural materials: they were untouched by man, except to shape them into their intended use. Knives, arrowheads, and even cooking tools were fashioned from simple materials.[101]

Then, before early humans began working in the fields, they learned to domesticate plants. Starting in areas with natural fields of grain, humans slowly learned to use new tools and technology to domesticate crops, making them easier to grow and process. Soon after, they cultivated their fields and built irrigation systems. Scholars now believe that agriculture was independently developed in at least 10 different places around the world. Places as far apart as Turkey, Syria, Southwest Asia, and China all experienced a transitional period to agrarian society, in which humans progressed from hunting and gathering to form settlements.[102]

Dated to the beginning of the agricultural period, archaeologists have found evidence for permanent settlements made of clay.[103] These early settlements used varied technologies in order to make sure their surroundings serve them well. Trees were cut and the forest was cleared to make room for the newcomers. Humans discovered that mixing water and clay could harden into material that could be used for the first permanent housing in history.[104]

These may sound like minor inventions to today's ears, but with this technological advancement, permanent settlements changed humans' social and familial lives. Most people cannot imagine their lives now without four walls and a decent ceiling. Certainly, they will not dare to marry and raise children without an appropriate roof to protect them.

Innovation did not just stop with sporadic settlements, though. New technologies were developed for hunting, mining, cultivation, and storage. Humans began to cluster together and develop ongoing relations with their neighbors. It was not clan members anymore who helped hunt a passing mammoth. Instead, social interactions evolved around cultivating yields, storing food in dedicated facilities, building housing to live in, and processing the ever-increasing variety of food they had.[105] Methods to fight insects and rodents were evolved and every technique was a real breakthrough that influenced diet for years to come. The Sumerians, for example, used sulfur compounds to control insects and mites, while early Chinese used chalk and wood ash for pest control in their homes and granaries.[106]

The Neolithic Revolution and its transition to settled agriculture soon expanded. Pastoral and hunter-gatherer societies found that the land they used for hunting was slowly being guarded by others due to the increasing overall population that meant more competition. As hunting opportunities decreased, humans were forced to turn to farming for survival, especially in times of climate change and population growth. A cycle began with one outcome: more and more people used the new technology of farming and settled down for longer periods of time. On the Yellow River in China, the Nile in Egypt, and the Ganges in India, river water was used to grow crops and agriculture spread quickly.[107, 108]

Natufians, for example, who developed their culture from around 12,000 to 9,000 BC, are credited with the development of the first farming tools. Sickles, picks, and other harvest tools have been found in their settlements. These tools granted a technological leap in harvest methods, adding to other development such as bows, arrows, and even jewelry.[109]

From these innovations, we can see an early accumulation of weapons. Although the weapons would initially have been mainly used for hunting, once humans settled and started to protect their borders and property, these tools were increasingly used for self-defensive and even offensive actions against rival tribes. Over time, cities with walls appeared, allowing for the defense of the population against other people and the wild. Settlements got larger and more prosperous, and agricultural technology slowly took over the former hunter-gatherer technology.[110]

As settlements grew, so did the farms around them and the domestication of animals was added to these new techniques. Humans bred pigs from wild boar, tamed goats, and started a great new friendship with dogs that served them well in keeping the sheep from going astray. Livestock advancements only added to a more productive, stationary lifestyle.[110]

These changes in technology affected demography dramatically. Humans at the beginning of the Neolithic period lived in communities of a few dozens to a few hundred people. As cultivation increased, the number of inhabitants grew exponentially. The global human population exploded from several million to almost a billion during this period.[111] But, agriculture not only changed population growth, it also changed human society itself with subsequent changes in family formation and relationships.

The Effect of Agricultural Technology on Society

Agriculture and domestication, as American historian Lewis Mumford writes, were "the first step towards capital accumulation."[112] Ever since the dawn of agricultural society, everything has centered on property. More than anything, humans started to center their lives around affluence and prosperity. Hunter-gatherers were free of this need because they needed to stay light in order to move quickly from one place to another. Food was an object to chase after, not to cultivate. The agricultural period changed this formula forever.

Once settled down, farmers were able to hoard property because they were stationed in one place. They cultivated their surroundings, created tools, and produced clothes to make sure they had enough not only for themselves, but also for trade purposes in case they wanted to barter those products against other needs they had.[104]

Moreover, the simple changes in technology created new fears to be addressed. If hunter-gatherers did not find food due to harsh conditions or draught, they could simply wander after their food. In contrast, farmers who settled down could not move so easily. Instead, they sought ways to protect themselves against times when climate conditions were not favorable.

Fears drive us all, even if we are not entirely aware of their effect. We sometimes make drastic moves to avoid conflicts, secure our future, or protect ourselves from the unknown. In exactly the same way, the new fears farmers experienced led them to be hostile toward their neighbors. They prepared themselves for harsh conditions by fabricating enough weapons and training the young in case they needed to go to war against more affluent neighbors.[113]

The more peaceful way of preparing for times of scarcity was to build granaries and accrue wealth and property. Preservation methods and storage techniques have developed over the years. These new developments allowed farmers to accumulate property of many kinds for longer periods of time.[107]

But even these developments created a social earthquake. While prehistoric societies tended to have very flat social structures because members of the clan worked together and focused on specific tasks, farming and the rise of property ownership changed this fundamentally. Once farmers started to accumulate wealth and increased their land ownership, the relative equality enjoyed by humans until then ended. Economic strata emerged, and there was a new need to protect the rights of those who had property against those who did not. The gap included real estate, luxury goods, or even the right to a particular livelihood. Over time, the population grew and land was harder to come by. Thus, being a landowner increasingly required great wealth.[47]

Empires like the Hittites and Egyptians emerged, where the elites lived off the payments made by local subjects. Rigid class structure and hierarchy complexes developed, and a general shortage of social mobility became central to agrarian society. In any given region, the ones who had the most property rose to power. Those who did not have enough land and resources worked in the fields and were paid by the owners for their services.[114]

To be fair, some leaders used the resources collected not only to fund their lifestyles, but also to provide services for local inhabitants. Technological advances were used to improve farming techniques with the aim to increase taxes collected by the rulers. This, in turn, helped society in general. The Romans, for example, brought with them superior techniques to benefit the farming community, such as land rotation, selective breeding, and new crops.[115] Later in that period, medieval feudal society had mechanisms in place to ensure that production continued. Here, too, local manorial lords exacted taxes and rents, but also provided services to the peasants, which, in turn, allowed aristocrats to carry on with their lifestyle.[116]

Consecutively, wealth accumulation and property ownership created the need to organize and monitor assets, a task which was resolved through another powerful innovation: writing. Property was carefully registered, inherited, and guarded through written records. Agrarian societies have been preoccupied with preserving family livelihoods and acquiring property, and with the rise of Sumer in Mesopotamia around 5,000 years ago, records began to be written down to organize who owns what.[7] The Chinese and Egyptians soon followed suit, and also developed written systems of their own at around the same time.[117]

Writing was a transformative invention. Whereas before, knowledge and stories could only be communicated verbally, now they could be written down and preserved for days, weeks, years, and often even longer. Knowledge could also be more easily transported all around the world. Literature was written to pass stories from one generation to another, while teaching children what is right and wrong. Financial transactions were recorded, and treaties were written.[118] All of this written material was key to preserving property rights and social structures.[119]

A widespread use of writing in the ancient world was found in Egypt.[120] Several types of texts have been found by Egyptologists, many of which date from the Old Kingdom (2700–2190 BC). While building pyramids and other projects, the Egyptians were copious record keepers.[121] Taxes were assessed against the population, orders from the king promulgated, payments recorded, and scientific knowledge transmitted. In short, the ancient Egyptians recorded, at a very early date, everything that was important in their society. So much so, that Egyptians are credited with inventing early accounting systems. Numbers recorded by scribes even include fractions, and quantities of up to a million can be found in hieroglyphs.

Alongside writing, the wheel was an invaluable invention; with it, commerce increased dramatically. Created around 3500 BC in Mesopotamia, the wheel first served for pottery creation, but was quickly turned to being used for carts. In this way, goods could be transported farther and more easily across geographies.[122]

Arising out of the idea of trade, the invention of money was perhaps the ultimate expression of the combination of writing, trade, and the transition to property ownership. Before this age, wealth was usually estimated in terms of land and possessions. Money changed this, and several nations began using currency. The first currencies appear to have consisted of shells and salt, as well as gold or silver. As metallurgy began to develop, so too did the materials from which coins were made; bronze, tin, and copper were all used at various times to create frequently stamped coins, after the example set by Alexander the Great.[123]

These coins soon became an international means of exchanging goods, which were being transported on the newly invented wheels. Humans now had easier means of recording, communicating, trading, and transporting—all essential components in making a functioning society that can interconnect with others and within itself. Systems were developed to deal with credit and debt within the merchant community and cheaters faced repercussions. For example, if a merchant defaulted on a debt, the local authorities would sometimes ban an entire nationality of merchants from trading in that area until the debt was paid. This caused a "peer pressure" form of debt collection and contract enforcement.[124]

Essential to this motive was the creation and monitoring of law. Laws were set up to regulate commercial affairs between the inhabitants of a settlement or country. First penned in ancient Babylon, the Code of Hammurabi is one of the first examples of codified law. Perhaps the most telling aspect of the code is that the largest section is about property. Hammurabi's laws included the very first commercial code and several sections covered credit, debt, and trading issues. Likewise, suing someone who has damaged your property and prosecuting a thief are procedures that date all the way to the time of Hammurabi's laws.[125]

It is easy to assume that everything is a matter of cultural differences and human development happened randomly and scattered. But it is amazing to see how, once some principles or technology spread, cultures followed very similar patterns of development. Researchers pointed out, for example, that the punishments prescribed by Hammurabi are similar to those found in the Hebrew Scriptures, which date from almost the same time period. For example, stealing can get you executed and if someone gouges out an eye, the punishment is that the offender's eye is removed.

A later example of agrarian-society law is the medieval merchant code. Applicable around Europe, this code was used to ensure that commerce was fair and cheaters were adequately punished. Essentially, the code was designed to solve the important problem of local-law differences. During the medieval period, local laws were not always binding on people who visited another country or were biased against those visitors. In response, the merchants wrote a sort of "code of conduct" for their trade.[126]

These sets of codes epitomize the rise of property ownership in that period. Between the general civil law codes like Hammurabi or the Torah and the medieval merchant codes, we can easily see how the acquisition and preservation of property were ensured. People who were caught lying, cheating, or stealing were punished with varying degrees of severity, including execution.

Another aspect of property ownership is that of humans owning other humans. Labor in agricultural societies was not always free. One of the earliest and most prominent examples of ancient slavery is the biblical case of Judaic people being enslaved by the Assyrians and then the Babylonians.[127] Likewise, ancient Athenians commonly had slaves. Most slaves in Greece helped out in their master's household, although there are reports of them being involved in business, and even travelling for their masters.[128]

After antiquity, a similar agrarian labor system developed: feudalism.[129] Unlike the ancient use of slavery, feudalism tied the poor people to the land and involved them in agriculture. The manorial lords would provide land and everything else the peasants needed. In return, the serfs had to pay taxes (and later, rents) to their feudal lord. Wealth flowed upward, where the people at the top were able to enjoy luxury and leisure.

To begin understanding how all of this leads to family and relationship formation, we should take a look at inheritance considerations. Closely connected to commercial sections was an extensive family law, which was important because it helped determined who owns what through inheritance. Throughout the agricultural period and all around the ancient world, medieval Europe, and elsewhere, inheritance had been an important topic in law and central to the agenda of governmental and religious authorities.[130]

Especially where families belonged to the agricultural class, it was important to have enough inheritance of land to earn a stable livelihood. After all, the preservation of agrarian societal structure required that property remains in large enough portions that agricultural activity be economically viable. This was an important consideration that led families to live together in multigenerational residences, pulling together land and labor resources.[131-133]

It is clear, then, that agricultural society was organized around the cultivation of fields and farms, and that property ownership stood at the center of its social dynamics. Social hierarchies sprang up, empires rose and fell, but there was one constant: social and economic efforts centered around the acquisition and preservation of property. Writing, law, and inheritance rules all sought to determine who owned what, and in what quantity. The multigenerational family was the perfect structure to fit these circumstances, as I show in the next section.

Relationships 2.0

Following the fundamental advances in agricultural technology, the meaning of family formation, marriage, and relationships changed markedly. The fundamental transformation from food gathering to food producing grounded people in one place. Instead of wandering, following the animals to hunt, humans built static houses and farms, and began developing pottery, jewelry, and fine stone tools. Because societies settled down, they did not need to bother with whether and how they would carry their accumulated possessions and a new notion emerged: home.

Marriage and relationships were deeply influenced by these changes. Whereas Society 1.0 focused on outbreeding needs and species survival, the sharp rise of property ownership necessitated measures to secure assets transference and preservation. Marriage, in this sense, helped to outline the boundaries of ownership and the ways new linkages to certain lands and farms can be created.

Marriage is not a static ritual; it changed even throughout this period and relationships took various forms. Yet, a major consideration in approaching such arrangements in agricultural society was to treat them as signifiers of property ownership. Belonging to a certain family means sharing in its wealth. This wealth was usually not of coins, which can be easily divided. Coins were not even invented at the beginning of the agricultural society. Instead, the main property included land and valuable items such as jewelry and weapons.

Thus, financial and property considerations were the basis of the decision to marry in all social strata. Among the privileged, relationships were contracted to combine landholdings and political power through endowments, patrimony, and social partnerships, and with the point of safeguarding bloodlines. Among the lower classes, simple survival required marriage, and men regularly picked spouses based on their potential economic value just as much as their reproductive capabilities. Working out in the fields required resilient women who could help with work, particularly during harvests. The fear of not having enough yield in the coming season ruled.

These considerations made sentimental love a luxury, at best. Who can think of love when rain is not coming, and there is a good chance that the coming spring will bring with it a disastrous yield? As inequality increased, marriage was an economic event, not a sentimental one. Marriage and birth circumstances could mean the difference between poverty and wealth: property would descend through a family, and social ties would be consolidated by marrying couples of equal monetary standing.

In Mesopotamia, we see an excellent example of the complex family customs that organized marriage as a financial transaction rather than a sentimental act. In *From Sumer to Babylon*, Jean-Jacques Glassner offers an in-depth analysis of family customs in Mesopotamia, emphasizing the importance of marriage in the lives of Sumerian and Babylonian people. According to his analysis, although men could technically make any woman of their choosing their bride by "coming into being" (with or without consent), most men preferred sealing the deal through a series of stages. The first one was, of course, choosing a bride. Families arranged marriages early for their daughters and, in doing so, ensured that for most of their lives, girls would be subjected to the authority of their in-laws. This authority over women is explained by seeing marriage as an extended issue of property ownership. A woman was bound to her husband's family and even agreed to marry his brother or another relative in the event of his death.

The next step of marriage in Mesopotamia involved making a payment called a "terhatum" and writing a legal bond called a "riksatum." Again, the exchange of money and the creation of a contract secured the young woman's fate as the future property of her husband.

Next came banquets and the gifting of food, called a "biblum." Families passed the "biblum" back and forth as a way of showing their alliance to each other as well as to show off their wealth. The family of the bride wanted to prove their daughter came from a desirable economic background, and the groom's family worked equally hard to confirm that their household could support a new member. The myth of Enlil, the ancient Mesopotamian god associated with wind, air, earth, and storms, and his wife Sud shows how important this gift-giving was. After Enlil sees the young girl Sud in the street in front of her mother's house, he assumes she is available because she is out in the street. But Sud refused to marry him, and Enlil decided to send her family gifts. Once her family received these gifts, Sud became his wife, was brought into his temple, and became mother-goddess, placing her in charge of all the secrets pertaining to women.[134]

In other cultures, marriage, as we know it today, was an even more distant concept. When one looks at texts from ancient Egypt, for example, it becomes apparent that early Egyptians have almost no word to describe the institution of marriage. The reason is that there were no official public or religious ceremonies to mark the event. For Egyptians, marriage was more about creating descendants for economic purposes, and less a legal or religious matter.[135] For this reason, Egyptian texts often used the phrase that a man "founded a household." Even later on, in contracts appearing around the first millennium BC, marriage was mainly decided by the husband and father-in-law, with the language stating that a father would "give" his daughter to the husband. A woman rarely had any say who she would marry.

The importance of property ownership in early empires continues with the requirement for a male heir to head the family after the father's death. In many cultures, if one wife could not produce a male heir, then men sought out another. Generally, a designated wife bore "legitimate" children who could inherit the family estate. Sometimes, men married priestesses who could not have children, so a secondary wife stepped in to take on the task of having a baby. We see this phenomenon in Babylon, Egypt, China, and India as well as among the Inca and Aztec people. As a rule, the sons of the wife would inherit everything.

Similar patterns persisted during the medieval period. Researchers show that legitimacy and illegitimacy were a world-wide concern for inheritance.[132] Wives and concubines were found frequently throughout the world, including Europe, Asia, and the Middle East. However, to inherit most property, a child had to be born to a particular woman prescribed to bear heirs for the man. This manifested itself in several ways, but the main one was that in most cultures restrictions on women were intended to protect them as an "asset": it was important that a man knew that the children from his wife were his and not someone else's.[133]

Thus, until the end of the period, marriage as a social institution was primarily transactional. Often what the bride and groom wanted was of little concern, and the marriage was a publicly mediated contract between their two families. Marriage was many times a necessity, not an ideal. Spouses were chosen shrewdly: the upper classes were concerned with keeping wealth or power in the family, and the lower classes were concerned with the need to earn a living. A marriage was successful insofar as it consolidated wealth and social status for both parties—the parties here being the larger families, not the individuals. Religious institutions would ratify the contract by blessing the marriage, which had effectively already been formed by mutual agreement, mainly based on economic reasons.

For this reason, secret marriages were a problem for the elite, because they made dynastic matches more difficult, and subject to the whims of potential spouses. No wonder that Henri II of France, who ruled in the sixteenth century, tried to convince the Catholic Church to require parental consent for marriages and to nullify those contracted without it. When the Catholic Church did not do what he wanted them to, it led to significant tensions. In the end, the Catholic Church decided to define the conditions for contracting a marriage much more strictly. Now, both the Church and the families had to consent to sacramental marriages. Instead of total autonomy, couples now had to depend on the priest.

It was not until the end of the Renaissance that marriage started to take its modern shape, as a form of companionship. Late-in-life marriage is an excellent indicator attesting this. Prior to the eighteenth century, marrying late was ridiculed by the community. After this point, however, it was considered acceptable to marry late or even to remarry in some societies, after the potential for children had passed, since it allowed for companionship.

Agrarian Technology and the Multigenerational Family

It is not only that Relationships 2.0 were mainly based on economic and property considerations, but also that they tended to be part of a certain family structure: the multigenerational household. With technological developments that brought people to focus on lands and farms, the multigenerational family was ideal. Blood ties signified property ownership, inheritance rules, and labor division.

Each member of the multigenerational family had their role in cultivating the land, working in the fields, and taking care of the livestock. Older people would live with their children and grandchildren, along with the rest of the household. In this way, everyone was taken care of in the family structure, and family property was key to family survival. In many cases, the members of such an extended family lived in simple houses with common spaces for cooking located close to their shared fields.[136]

In turn, the family became a group that lived together for a common purpose, such as running a farm or business. The group might include apprentices, foundlings, servants, and slaves. In some extended families, "surplus" children often became the underclass, working as servants, hired farm hands, and apprentices. They might also go into the army, or work for the local landlord. Other extended families annexed disinherited children from other families to live with them as servants. Throughout the Mediterranean, for example, we see them counted with their masters' wives and children in censuses.[136] Thus, the basic unit of society turned to be the multigenerational family that consisted of parents, children, grandparents, and other close kinship. Across agricultural cultures, multigenerational co-residence was the most common form of living.[137]

This notion of the family as a multigenerational household rather than the nuclear family lasted until industrialization. Perhaps the simplest way of seeing this is to look in the dictionary. Dictionaries in the sixteenth and seventeenth centuries define the word "family" as an entire household, which included extended family and even the servants. By the eighteenth century, dictionaries were starting to see the family in narrower terms. This change coincides exactly with the transition from agricultural society to industrial society in Europe.[138]

Indeed, it took centuries before the romantic model of marriage became the rule and the nuclear family emerged. It was not until the late eighteenth century that it was more popular and acceptable to choose a partner based on affection rather than economics. Mainly beginning with the lower classes, interference with spousal selection was reduced and romance became a bigger factor. Among upper classes, parental input in spouse selection was still economically motivated, at least for another century.[139] Since upper-class parents were in a position to provide financial support, they used marriage to improve commercial and social prospects and keep wealth from being dispersed.[138] This major shift toward romantic relationships, which again coincided with major technological developments, is explained in detail in the next chapter.

3
Relationships 3.0

As human history advances, the periods are getting shorter, and changes are appearing more often. A Roman born in 200 BC and a Roman born in AD 200 shared very similar worlds. But a fourteenth-century English cavalier and an eighteenth-century Englishman working in a Mancunian factory would not recognize each other's way of life. Stability was the rule in past times. But this rule changed in the eighteenth century, along with seemingly everything else.

Marked from the mid-eighteenth century until the mid-twentieth century, the Industrial Revolution fundamentally changed everyday life in most places. After more than 10,000 years of Society 2.0, in which humans settled down, farmed their lands, and harvested their yields, technology made another leap. Inventions such as steam power and electricity changed humans' lives drastically as factories producing goods with overwhelming speed and efficiency began to spread around the world.[140, 141]

There was a curious turn in industrialized society: while humans had domesticated their surroundings during the agricultural revolution, now factories and manufacturing processes started to "domesticate" humans. New machines and production methods became centers around which people raised their families and lived their lives.[142, 143] As time went by, humans flocked to cities to work in factories,[141] and shifted into a world where technological developments became the center of society.[144]

In turn, the industrial society carried significant implications for marriage and relationships. Although the mechanisms at play are complex and not entirely uniform, evidence suggests that industrialization contributed immensely to the rise of the nuclear family. The nuclear family that better suited urbanization and industrialization increasingly replaced the multigenerational family that surrounded fields, farms, and crops in the past.[145, 146, 147] In an agricultural society, large families shared lands and livestock; hence a major reason for marriage arrangements was the need to preserve property ownership. In contrast, families in industrial society were centered around the new means of production, lived in scanty city houses, and thus needed to nuclearize.

Technology and Industrial Society

Even before industrialization spread, changes in the production of knowledge augured this period. In the early seventeenth century, Francis Bacon advocated for systematizing how we acquired knowledge through data and measurement.[148] Our modern scientific method—the idea that there is a set of agreed-upon means of acquiring knowledge—has hardly changed since its genesis with Bacon. Only now, as artificial intelligence is shifting the way we discover things from observation and reasoning to data mining and de facto answers, have we become fully aware of the importance of Bacon's revolution in the history of human thinking.[149]

The same goes with later publications, such as Darwin's *On the Origin of Species* from the nineteenth century. This and other publications changed the way we think and discover new things in our world. Humans began to trust their examinations rather than their beliefs and thus brought about a technological leap. Science and scientific investigation in the modern sense began to grow and gain more attention.[150]

Connected to the change in producing knowledge was the shift in disseminating that knowledge. The advent of mass printing and the spread of technical schools were crucial in this sense. While knowledge is valuable in and of itself, keeping it in monasteries and the sporadic universities that existed until that time did not serve science well. Without disseminating the diagrams, discoveries, and models that had developed up until that point, others could not advance learning. Rotary printing on long continuous rolls of paper made ideas more easily accessible and helped catch the eyes of brilliant minds that were unknown before.[142] The spread of technical schools in the nineteenth century further encouraged the diffusion of scientific knowledge. These new schools provided another opportunity for science to develop and sparked a new wave of innovation.[151]

Since the beginning of the Industrial Revolution, the world has witnessed more incredible feats of innovation than at any other time in human history: the mechanical weaver, the sewing machine, new iron processing methods, steam power, the production of new chemicals and cement, the telegraph, dynamite, the modern assembly line, and so many more.[152] All of these inventions made the production of goods more efficient. They stream-lined the tasks required for survival and paved the way for even more significant inventions.[153]

The immediate effect was on food supply. Pre-industrial societies were plagued by scarcity, stemming from the low productivity of labor. Industrial society solved this problem. Through the mass production of goods and the development of fast and efficient transportation systems, industrialization changed the face of agriculture and the food industry.[154] Due to new technologies, humans were able to access an adequate supply of goods, and the population grew dramatically, especially in later phases of that period.[155]

Deep impacts were also registered in the very notions of time and space. With gas lighting and then the light bulb's invention, dark corners were made into mental and physical spaces for work and living at nighttime. Light quickly spread through cities and towns across the world and revolutionized the meaning of day and night. It allowed factories to continue working long hours and improve productivity drastically.[156] Similarly, steam and electricity were perhaps the most critical prerequisites for changing the idea of space. Faraway places became closer with railway transport, and traveling time was reduced.[157]

Technology, Religion, and Relationships

In many ways, these technological changes had a ripple effect on religion, which, in turn, affected relationships. Until the eighteenth century, societal rules were dictated by the Church and enforced by the State. Bad behavior, as defined by the Church, was often exposed and mocked in the public square. The state executed spiritual dictums, and constables could investigate suspected blasphemy, even to the point of breaking into a house. Men were considered the household's spiritual and legal heads, and it was necessary to enforce this authority by any means necessary. No wonder, then, that people were bound by duty and obligation rather than loving feelings or self-interest. The Church tended to discourage couples from enjoying sexual intimacy, treating it more as a duty necessary for procreation.[158]

The weakening hegemony of the Church, especially in Europe and its colonies, marked a significant change in this sense. Although complex and not wholly universal, the technological revolution had a crucial role in this shift. Evidence shows, for example, that scientific curricula clashed with religious dogma.[159] In addition, science brought to the public's attention discoveries that undermined the doctrines expounded by priests and pastors.

Thus, empirical observations replaced old axioms, science weakened trust in old traditions, and life became oriented around the new means of production. In particular, the Enlightenment movement, characterized by an intellectual enthusiasm and unruly skepticism and atheism, lay the foundations for the crisis of faith. Slowly, thinkers unveiled new concepts of social order that did not require the Church. In turn, religious influences on society began to weaken in the eighteenth century. It was Nietzsche who outrightly declared "God is dead. God remains dead. And we have killed him."[160]

As the Church and moral crusaders lost control of society, religious opinions became less important. If something was not a threat to the political order, it was more likely to be a private matter that the State and the Church have nothing to do with. This was a time when young people had relatively less pressure to conform to social norms, while civil authorities loosened their grip on citizens' lives.

Thus, as privacy became more important, the nuclear family rose to prominence. This was simply because certain aspects of life became private matters for those in close relationships to decide for themselves. People started emphasizing relationships between individuals, and formal marriages on a sentimental basis became more common. In turn, the view of couples being bound together by affection rather than by economic consideration quickly spread.[161, 162]

Interestingly, these changes led to a second phase in some countries, in which couplehood and the nuclear family were actually in jeopardy. As the old norms eroded, lower-class people, in particular, were more likely to do whatever they wanted sexually. No longer needing Church approval for their actions, people were less inhibited sexually and jettisoned old moral behavior norms. As a result, illegitimacy increased, and marriage itself declined.

Yet, the Reformation, and in turn, the Catholic response, fought this change in the view of marriage. Cleverly, they turned relationships and marriage from a transactional "necessary evil" to an important social institution. Marriage became a holy institution, and the Church emphasized that marriage helps keep society together. From then on, there was a religious reason for Catholic teaching to encourage matrimony, and the push toward Relationships 3.0 grew stronger.[163]

Perhaps more than anything, the story of Relationships 3.0 is the story of the industrial city. There is no place the radical changes of industrialization can be seen more clearly than in the rise of the industrial city, and there is no place better than the city to see how relationships and marriage transformed.

The Rise of the Industrial City

Industrialization and the rise of big cities went hand in hand, transforming rural, agrarian societies into urban, industrialized ones. Here, smoky skies, busy streets, and the anonymous crowd became home to a new generation that cut ties with traditions, religions, even history. The new, the modern, and the secular thrived, and the multi-generational family became an artifact of times past. Children of multi-generational families that were once surrounded by their fields, grew up and moved to cities to set their lives around big factories. Men migrated to towns with their wives and children only. There was no need for the whole family any longer, and the industrialized city embraced a rising phenomenon: the nuclear family.

The shift from agrarian farms to urban centers happened inexorably. New agricultural methods allowed for increased land productivity. Inventions such as the horse-drawn seed press and chemical fertilization made farming more productive and less labor-intensive. Thus, the local, self-sustaining, and decentralized economies transformed into commercial agriculture.[164] Goods that had been hand-crafted in the past were produced in large quantities, making many farmers redundant. In turn, farmers moved to the city in search of opportunity. They abandoned traditional agricultural practices that had been the primary source of sustenance for ages and adapted to the fast-moving urbanization and industrialization.

The city, on its side, became more appealing. Spreading transportation networks made the economy of the industrial city more connected and reliable. Before, towns were at the mercy of local weather and local crop yields; if drought struck, they had to pray their food stores would last them until the next harvest. But with fast and efficient transportation networks, food surpluses in one place could easily compensate for shortages in another.

The coming to life of the industrial society also resulted in an increased demand for labor in factories. The thriving factories in the cities appealed to rural people who were eager to improve their social status. The example of John O'Neill, an Irish man who documented his life, is typical.[165] O'Neill traveled from Cork to London in 1808 searching for work. Upon arrival, he found a job in an Irish workshop in the city. When he earned enough money, rather than return to Cork, he sent for his wife and children and moved permanently to London. For many like O'Neill, the work originally sought was seasonal but quickly became permanent, leading to permanent residence in the city.[142]

In addition, the city was flooded with luxurious products. Farmers, who witnessed an agricultural output that grew faster than the population, began to consider exporting more of their goods to the city. As productivity increased, they started to expand beyond just food and sustenance, growing more sugar, tobacco, and cotton. These crops had little or no nutritional value but could be sold to the cities for a handsome profit. In this way, the city was filled with lucrative goods that drew many.

No wonder, then, that individuals and couples migrated from rural areas into the big cities to seek employment opportunities and quality of life. In England, as a prominent example, the share of the population living in cities jumped from 17% in the early nineteenth century to 72% toward the end of the century.[104]

While these changes characterized European cities at the beginning, developing countries were quick to follow. Especially with the spread of colonialism, the concept of the city as a hub of modernization and the center of industrialization was duplicated in other countries such as Mexico, India, China, and several countries in North Africa and the Middle East.[166] From the sixteenth century onward, urban conditions transformed in many developing countries: in Latin America, old structures were destroyed, and new buildings were erected by the Spanish and Portuguese; in Africa, major settlements such as Johannesburg, Cape Town, and Nairobi were established by the French and the British; Calcutta, Bombay, and Madras followed suit in India; and even in China, North Africa, and parts of the Middle East, where native centers remained, the influence of the West was felt, and urbanism took center stage. It is clear how, throughout the world, the city rose to power, leaving behind agrarian society.[167, 168]

Life in the Industrial City

To understand why urbanism had such a tremendous impact on family and marriage, as I will later show in detail, one must realize city life's nature and how it led to social fragmentation. Living in the city took its own character and influenced society as a whole, as more people flocked to metropolitan areas. Collectivity and social control weakened in cities more than in villages, and the formation of social networks was harder in the urban environment. City life was more individualistic as a result, and instead of collective identities, the construction of personal identities became necessary.

Three major aspects of city life can explain this process of social fragmentation: social inequality, work compartmentalization, and housing arrangement. First, as the accumulation of wealth progressed, so did the establishment of social classes.[169] The industrial society prompted the emergence of stratification, in which different social classes diverge sharply. While some individuals acquired wealth, others lived in abject poverty. Historian Lawrence Stone likened the economic strata of the time as "a huge hill with a high and extremely thin tower on it, with the bulky hill consisting of the mass of poor peasants, artisans, and laborers, and the thin tower consisting of the courts, the large landed aristocrats, and the rich merchants."[170] Although some argue that the middle class became larger during that time period, most agree that stratification increased and the poor became poorer.[171]

The owners of businesses and factories were those who mostly occupied the upper classes of industrial society. This was the new type of property. They included ship owners, industrialists, and merchants who had accumulated considerable amounts of wealth and capital. A thin layer of middle-class merchants, lawyers, and doctors also enjoyed the advantages of the new prosperity brought about by industrialization. Many of these "capitalists" or the "bourgeoisie" began spending their recreational time entertaining themselves in concert halls, theatres, and sports facilities. While the poor were forced to remain in the cities to work and survive, the affluent could afford to reside at places where they could avoid dirt and disease while still taking advantage of the city's benefits of learning and power. As Richard Cantillon, an Irish-French economist from the eighteenth century, recounts: "the landowners who reside in the provinces do not fail to spend time [in the cities] and to send their children there to be polished."[172]

In contrast, the majority of city dwellers, the members of the working classes, were forced to live in unhygienic, overcrowded places and were ill-treated in factories. Poor diet, overcrowding, inadequate sanitation, and medieval medical solutions all contributed to the public's poor health and lower life expectancy. The ill-constructed, unplanned, and densely packed working-class neighborhoods enhanced the spread of deadly diseases, including tuberculosis, cholera, typhoid, influenza, and typhus.[170]

The class divide was especially pronounced in the factories, which lacked any safety features and went to extreme measures to keep costs low, including employing orphans. Early photos of factory workers often depict weary men in ragged, filthy clothing, and one or two fingers conspicuously absent. Factories remained dangerous places to work for centuries, and it was not uncommon for employees, adults and children alike, to lose fingers or hands on the job. What is more, many children, out of necessity, actually sought out factory work to help support their families. Frances Trollope, an English novelist from the nineteenth century, described this reality in her book:

> A little girl about seven years old, whose job as scavenger, was to collect incessantly from the factory floor, the flying fragments of cotton that might impede the work . . . while the hissing machinery passed over her, and when this is skillfully done, and the head, body, and the outstretched limbs carefully glued to the floor, the steady moving, but threatening mass, may pass and repass over the dizzy head and trembling body without touching it. But accidents frequently occur; and many are the flaxen locks, rudely torn from infant heads, in the process.[173]

The second aspect of city life in industrial society was the compartmentalization of work. Industrial society was driven by mass production technology of goods to support its large and increasing population. As cities industrialized and factories filled the skies with black smoke, the lives of those who worked in factories increasingly revolved around their workplaces. Machines became not only the solution, but also the sun around which we orbited.

Humans needed to mind these machines, and they quickly became appendages to these new technologies rather than the other way around.[172]

With industrialization and its efficient manufacturing processes, workers became more replaceable. On a manufacturing assembly line, each worker had a very specific job, and this job was so modular that another worker could easily step in and learn what needed to be done. Hence, the idea of a working lineage or family occupations—fathers teaching sons highly skilled, apprenticed trades—became obsolete. Traditions of tailoring, shoemaking, milking, or simply knowing the family land intimately became old-fashioned. Sophistication, efficiency, and state-of-the-art technology were the new names of the game. Industrial society focused on using non-human sources of energy to increase the scale and rate of production. There was little room left, therefore, for caring about people and maintaining traditions.

The third aspect of city life in industrial society was housing arrangement. As people moved into the cities, seeking employment and escaping the uncertainties of agricultural life, the cities faced a rapid population increase that required immediate housing solutions. Most cities were ill-prepared to handle this shift. Used to smaller populations of tradesman, merchants, and a handful of laborers, the addition of so many unskilled laborers or semi-skilled tradesman created an unbearable burden on urban development in many cases.[142]

We can get a glimpse into that world by looking at the recent history of China. In cities such as Shanghai, Ningbo, and Guangzhou, massive factories filled with workers produce large quantities of goods for the world. The workers live close to the factories, as close as possible, in tiny living spaces to reduce congestion. Overcrowding is still a major concern for China as its industrialization is ongoing, and these conditions have only started to relax recently. In 1978 there were still merely 3.6 square meters of living space in urban housing per person. By 1985, the living area had increased to 6.7 square meters per person. Only recently has there been a significant improvement in living conditions.[167, 174]

Like China in recent history, policymakers and city planners in many countries throughout the industrial period desperately tried to respond to labor's rapid growth. Cities filled with housing options based on short-term solutions, but they proved less than ideal. Row buildings, hastily constructed of wood, often popped up to house the underprivileged employees. These offhand solutions for workers remained decidedly primitive for many decades. Such buildings were poorly built and usually crammed with workers, making for an unhygienic environment. Disease and pollution became rampant in early cities, as workers jam-packed into whatever housing they could afford.[142]

In parallel, early industrial stages required close proximity of workers to the factories due to lack of reliable transportation networks and demand for long working hours. As early cities lacked convenient methods of transportation—except horses, which the average worker could not afford—such housing was often located close to the factories and workshops. Families often lived in single rooms in cheaply constructed buildings packed as close together as possible to allow for the maximum number of homes.

Such houses lacked any plumbing or washing areas. Personal washing needed to be done outside or using a tin tub with water drawn from a nearby well, with an outhouse serving more regular toiletry needs. One striking evidence for this condition is a pamphlet published in 1883 that caused a parliament sensation, titled "The Bitter Cry of Outcast London." Its author wrote: "Walls and ceilings are black with the accretions of filth which have gathered upon them through long years of neglect. It is exuding through cracks in the boards overhead; it is running down the walls; it is everywhere."[175]

In countries of immigration, such as the United States, immigrant groups often banded together for protection and familiarity, creating neighborhoods named for those who resided there. Examples include Chinatown in New York, Little Italy in Baltimore, and Czech California in Chicago, to name a few. These ethnic groups created their own churches, storefronts, and social clubs. African American populations in the United States were often placed in even more rundown, cramped living quarters. Their neighborhoods were carefully separated from whites.[176]

These three aspects of the city—social inequality, work compartmentalization, and cramped housing—created a mental and physical fragmentation that grew into a family nuclearization, a process that gave rise to relationships 3.0.

Relationships 3.0

In light of this social reality, it is clearer how marriage and family changes came about during industrialization. With social fragmentation, people felt more isolated in the city crowd. In turn, there was an increased emphasis on being happy within one's home. Without the village, the collective work, and the interaction with others, people turn inward into their homes and nuclear families. Social fragmentation in the city made this arrangement more desirable, and the nuclear family became one of the only mental spaces left with support.

Thus, the home became more idealized and a new, pronounced distinction between public and domestic life emerged. By the end of the nineteenth century, the working classes had fully adopted this new mentality of "separate spheres" and saw home as a welcome retreat from the bustle and hum of the city's factories. Such a distinction would have made no sense to the farmers of agricultural society, where work and home life happened in the same place. Moreover, while rural life allowed the large multigenerational families to live in one home, and in many cases, required such a structure to survive, the cramped confines of urban life prohibited such developments. Small apartments had no room for multigenerational family members. Forced into ever-smaller rooms of aging buildings, divided repeatedly to house a rapidly increasing population, families needed to change their familial structure to adapt to urban life.[142]

Even more, as families matured and separated, what bound multigenerational families together gave way.[172] Those who worked in the factories slowly developed an identity separated from the collective identities of their old communities. It was often the young only who ventured to the cities, leaving behind their parents and extended family. There was less need for the multigenerational family in factories.[177] Once couples started to live away from their parents and relatives older generations had less power over their offspring, and the young sought companionship and emotional connection with their partner.

In these ways, the nuclear family rose to prominence in the eighteenth century, starting in Great Britain and spreading to other parts of the world. Many sociologists, demographers, and historians suggest that the notion of "family" as a close-knit group of relatives was weak until the sixteenth and the seventeenth centuries. Prior to the seventeenth century, marriage was more about the survival of communities, raising children for the lower classes, and preserving wealth for the upper classes.[178]

The degree to which this was true varied depending on the region. Family nuclearization following industrialization fits Scandinavian history particularly well, for example. The same shift was arguably less pronounced in England, since the close-knit nuclear family existed slightly earlier.[179] Still, as a general trend, the nuclear family's consolidation in this period holds across geographies, despite complex intervening mechanisms. This is also generally true in Asian countries such as Japan, Korea, and China. The only difference is that industrialization arrived later. But, once it began, a transition from multigenerational to nuclear families was apparent.[177, 180]

The transition from multigenerational families coincided with marriages becoming more affectionate. As the economic importance of marriage declined, spouses became lovers, companions, and partners. The structural changes in family formation altered the very ways that family members thought and felt about each other. It is unclear when the shift to a more "sentimental" approach to family took place exactly, and in what order. Some scholars see the changes happening across geographies more or less uniformly over time, while others see more regional variation. But, however these cultural shifts unfolded, we see families becoming more affectionate, women gaining greater equality in marriage, and children receiving more nurturing from their parents.[178] For the first time in such broad adoption, feelings and romantic sentiments became the basis of family ties.[181]

Of course, it would be too simplistic to say that the move to the city and the social organization around the factory, following industrialization, were the only forces shaping the nuclear family. But it is clear that the technological development at that time seeped into family structuring through several cracks and crevasses. Evidence to support this transition after industrialization, although complex and composite, is plentiful.

The first scholar who convincingly argued that such a transition occurred is Frederic Le Play, who lived and published in the nineteenth century and knew the period almost firsthand. In his study of 132 families in different European, North American, Asian, and North African countries, Le Play showed that agrarian societies tended to be composed of multigenerational, patriarchal families. In contrast, the shift to industrialized society came with the nuclearization of the family.[182]

Le Play showed that only the oldest son remained with his father (stem family) in these families, while other children spread around and set new families. Thus, couples in industrializing regions were increasingly less likely to live in conjoint, multigenerational households. Although some multigenerational arrangements continued in some places after industrialization, and the nuclear family partly existed before industrialization, this was the rule. Interestingly, Le Play actually criticized the nuclear family as immoral, an attitude that shows how new this concept was at the time.[161]

The founders of sociology, such as Durkheim, Ferdinand, Georg Simmel, and Max Weber, quickly joined Le Play. For example, Durkheim argued that social differentiation in urbanized areas led to the loss of the multigenerational family structure. Specialized functions of the extended family, which were customary in the village, weakened with urbaniztion.[147, 183] Others continued with this line of interpretation. By the mid-twentieth century, many agreed that urbanization and industrialization resulted in nuclear families adapting to the new circumstances.[184]

Theorists from the early twentieth century explained this relation in the "theory of social breakdown." They had a twofold argument, explained earlier in this chapter: first, the nuclear family fits the requirements of industrial life better than multigenerational families due to new housing and working circumstances; second, migration from the village to the city disintegrated families and led to the erosion of kinship networks.[185]

The many scholars who advanced the argument regarding the connection between industrialization and family nuclearization were part of industrialized society and experienced these changes almost firsthand. So, it is surprising that beginning in the 1960s, historians started a fierce attack on their interpretation, an attack that is still ongoing.

Yet, a growing body of literature shows that this attack is unjustified. The methods and measurements of those claiming there is no connection between industrialization and family nuclearization were called into question.

For one, those who argue the nuclear family was pervasive before the Industrial Revolution used cross-sectional records. This means that they took a one-time snapshot from reality at that time. However, families and households evolve over time and so are better suited to study by longitudinal analysis.[137, 146, 161]

One example of such a comprehensive, longitudinal analysis was conducted on data from Le Creusot, a village in France that turned into an industrialized city between 1836 and 1886. The researchers used data from 10 linked censuses in five-year intervals to analyze family patterns over these 50 years. They show how multigenerational co-residence declined. Interestingly, this phenomenon happened despite policies that encouraged people to stay in a multigenerational family structure. These policies included sanctioning access to labor, housing, and schooling. Nevertheless, people still formed nuclear families instead of extended ones. Industrialization and rural-to-urban migration drove family formation in the opposite direction. They were forces too strong to resist.[186]

Life expectancy and other demographic characteristics at the end of the agrarian period might also explain why fewer multigenerational families were found in the past than expected. Simply put, the combination of lower life expectancy, late first marriages, and delayed childrearing reduced the probability that more than two generations of family members would live under the same roof. However, this does not mean multigenerational families were not the norm before industrialization. Had the parents lived at the time their children married, the expectation was that they moved in with them.[161]

A recent study compared pre-industrial towns with industrialized cities in Germany in the first half of the eighteenth century.[162] Using census data from that period, which is considered highly reliable, this study concludes that the industrialization of cities was "the final victory of the nuclear family."[162]

Steven Ruggles from the University of Minnesota is an authority in demographic studies. He is the director of the Institute for Social Research and Data Innovation and the creator of IPUMS, the world's largest population database. Ruggles brings substantive evidence supporting the conclusion that urbanization and industrialization prompted a transition from multigenerational to nuclear families.

Ruggles analyzed an impressive collection of newly available data from 87 censuses of 34 countries between 1850 and 2007. He showed how, for example, the percentage of elderly living in three-generational housing in the United States plummeted from around 30% in the mid-nineteenth century to approximately 15% in the mid-twentieth century. He argues that the rise of industrial labor opportunities encouraged the younger generation to leave the farm. The decline of agricultural employment among the younger generation was a key variable in the drop of multigenerational co-residency in the United States.[147] Ruggles states:

Growing commercialization and industrialization in Northwest Europe and North America in the nineteenth and twentieth centuries meant that a declining percentage of families had farms. Young people moved to towns, attracted by the high wages and independence offered by jobs in commerce, manufacturing, and transportation. Thus, economic development undermined the material incentives for multigenerational co-residence, and gradually the elderly began to reside separately from their descendants.[147]

Ruggles is decisive in his opinion that industrialization and technological development have impacted family formation fundamentally and contributed to the creation of the nuclear family. This new family structure lasted throughout the industrialization period and well into the twentieth century. The separation between the spheres was breached only when another technological revolution came into force: that of information.

4
Relationships 4.0

Hosni Mubarak, the President of Egypt for 29 years, thought it was just another demonstration, one that he could brutally halt in its tracks like so many before. This time, it began with the "We Are All Khaled Said" Facebook campaign, publicizing the death of a young man by the police for not showing his identification card quickly enough. Twenty thousand people went to the streets in response and flooded Tahrir Square at the heart of Cairo. News spread exceptionally quickly, and protestors were mobilized thanks to the use of online platforms. They protested governmental policies of violence, abuse, and torture. While the whole world observed the uprising with a mix of astonishment and disbelief, the large turnout inspired further protests that eventually unseated the regime. In February 11, 2011, only 18 days after the demonstrations began, Mubarak's regime ended, and he soon found himself in jail. Mubarak knew he had made one major mistake this time: he ignored the power of information.[187]

Since the mid-twentieth century, when modern computers started to spread, information has flooded society. Information connected people, organizations, and cultures and helped them grow and improve. Countries became globalized, corporations turned multinational, and societies evolved to be diverse and multicultural. Knowledge, rather than material goods, moved to the forefront of the economy, and the service sector started to generate more wealth than the manufacturing sector. Markets saw value in knowledge and started to put a premium on information. In turn, many industries began to focus on accumulating, analyzing, and disseminating data.[188]

Today, even the smallest piece of data has the potential to power a butterfly effect–style change on the rest of society, creating another node on the web of social forces. We are constantly bombarded with information as we tackle the previously inconceivable power of news and "fake news." Almost every day, someone from any side of the political aisle—the far-right or the far-left, Fox News or CNN—accuses the other side of misusing information to gain control over politics, public opinion, and resources.[189]

While this sounds like a matter of politics or communication studies, the issue relates very much to the family in the Post-Industrial Age. Information has more power than ever before, and it has shaped all aspects of society, including our love lives. Whereas a hierarchical class and centralized power characterized industrial society, the exponential rise in information has made contemporary society multicentered and networked.[190] We have evolved into a society of individuals with multiple nodes on a vast and global social network, on which each individual is a center for various social forces.[191] The new networked structure further atomized relationships. Within a few decades, the marital dream collapsed, and the short, Golden Age of the nuclear family found itself quickly replaced by the solo generation.[192–194]

This chapter, therefore, shows how relationships and kinships, which have progressed throughout human history, reached a new level in the recent information age. As discussed in previous chapters, clans formed the base of hunter-gatherer society, multigenerational families built agricultural society, and the nuclear family rose to prominence in industrial society. In the information society, however, the individual stands as the focal point of human society. Television, computers, the Internet, and smartphones brought a process of atomization and individualization to society. As every piece of information became another node on the web of knowledge, every individual became a vector on the social web. The family, in turn, saw a sharp decline in popularity. Instead of the family being essential to breadwinning, procreation, and social status, individualism rose to power, and the values of freedom, creativity, and trying new things became central.[195]

The Rise of the Information Age

The information age is marked from the mid-twentieth century until around the 2010s, although it continues today and overlaps with Society 5.0.[10, 196] Obviously, we do not change all at once and we carry our previous habits and behavior patterns with us, as we did throughout human history. The information society emerged in the post–World War II era, when increased interest in storing, retrieving, manipulating, transmitting, and receiving information in a digital form spread among mathematicians, engineers, and scientists. They believed that such technologies would provide a crucial step toward realizing automation's dream to improve existing communication, processing, and manufacturing methods. Thus, a world that had once been primarily dependent on mechanical means became more digitally based.[197]

Although there were many breakthroughs introduced in recent decades, three innovations stood at the center of information society: computers, the Internet, and wireless phones. Computer technology was developed after the Second World War with one revolutionary idea: information can be automated. The assumption was that significant technological and economic progress could be achieved by building up digital machines that not only enhance the accuracy and efficiency of calculations, but also fully automate them.[198]

In this way, the amount of information increased exponentially and became the center of society and economy. The year 1977 is perhaps when computers truly achieved that goal. The TRS-80 Microcomputer System, Apple II, Commodore PET, and Compucolor II were all released in 1977, creating the market for affordable home computers. Personal computers quickly became popular, and society was flooded with information that was more accessible and better handled than ever before.[199]

It is striking to see how access to information and the ability to extract, use, and disseminate information became the dividing factor between successful and unsuccessful industries with the emergence of this era. Sophisticated computerization made high-volume mass production more efficient and affordable. The computer allowed manufacturers to measure everything accurately and to quickly fulfill demand. The monetary value of information and information-related industries increased exponentially as access to information became a powerful means of growth and prosperity.[200]

The information age expanded further with the development of the Internet, whose popularization and commercialization made information more accessible and visible to people throughout the world. The late 1960s saw the first network dedicated to scientific communication. The Internet then was structured to only deal with data transference, and the only extension was to the American military in 1975.[201]

Yet, around the early 1980s the potential of the Internet to enhance public communication emerged, thanks to expanded access, with easy-to-use software, and regulatory changes that made online communication simpler. Graphical interfaces and increasing commercial investments helped the Internet to spread more quickly, and the 1990s saw a significant increase in Internet usage in many countries.[202]

In the 2000s, with fast and consistent technological advancements, the Internet became a hub of social, political, and economic activities—a crucial element for human existence and, to an extent, economic growth. Amazon, Google, Alibaba, Meta, and Netflix emerged as economic titans and trendsetters that are housed on the Internet. Today, every single minute sees around 4 million Google queries processed, 4.5 million YouTube clips viewed, and 40 million messages sent via Facebook or WhatsApp.[203]

Together with computers and the Internet, the third factor involved in the expansion of information was wireless technology. Cell phones brought ease and frequency of communicative exchanges to our fingertips. In 1992, with the introduction of the first true smartphone by IBM, the cell phone developed even further and turned into a handy computer.[204] It took several years until smartphones were widely available, but the very idea that not only other persons, but also information itself can be accessed and retrieved everywhere was revolutionary. From then on, computers, the Internet, and wireless technology have been integrated into one device that sticks information and people together.

Long-distance communication and information no longer relied on telephone lines or stationary machines, with 2002 marking the first year in which the number of wireless subscribers exceeded fixed-line subscribers throughout the world. While only 16 million wireless phone subscriptions were active in 1991, this number exceeded a whopping 4 billion two decades later, meaning the majority of the world's population have had access to wireless communication since the 2010s. Today, there are more wireless phone subscriptions in developed countries than people, with some people having two and three subscriptions in their name. Even poor countries give access to communication mediums a high priority despite financial hardship.[205]

Three key players pushed these technological advancements forward: economists, policymakers, and scientists. Economists were looking for a way to enhance productivity gains through automated systems rather than manufacturing processes. The steam engine, then electricity, towered at the center of the industrial society's technological advancement. Both amplified and substituted for human *physical* labor. In contrast, the function of the new, innovative technologies was to amplify and substitute for human *mental* labor.[206] If everything is measured, calculated, and operated by machines, economists argued, production processes will be cheaper and more effective. Moreover, they continued, new knowledge should be embedded in processes, organizations, and products. This would be an important driver for economic growth by enhancing the ability to innovate and invent new products with high efficiency.[207]

Policymakers interested in maintaining growth soon joined economists. Governmental and semi-governmental experts analyzed and gauged policy measures and compared them across countries. The assumption was that access to data and collection of information should be an essential principle in post-industrial policymaking. Three aspects were employed in this sense: the first was data collection, analysis, and dissemination; the second was setting quantitative performance measures and outcome-focused goals; and the third was developing systems among local and national institutes to ensure that incoming information was used all the time to guide policy priorities and solutions. Over time, information became inseparable from policymaking processes, on all levels.[208]

On their end, engineers and scientists in the information sector worked to increase access to knowledge by creating better systems for collecting, analyzing, and disseminating information.[206] They developed better infrastructure and more sophisticated methods to handle data. In turn, the volume of information available through myriad databases expanded exponentially. This infrastructural revolution has made it possible for various industries and organizations to function more accurately and efficiently. No wonder, then, that global investment in data centers alone is around 200 billion dollars each year in recent years.[209]

Working together, these players propelled the development of better and more advanced information systems. These systems became valued in the information age because they have made life easier for humans through improved efficiency and profitability of businesses, organizations, and governing bodies around the world. They, in turn, became an integral part of daily human life, and gave birth to the information-based economy, culture, society, and politics.

In the economy, the result of these developments was that information-related industries flourished and became the growth driver. Financial institutes became the production and distribution centers for information goods such as credit analysis, value estimations, and trading data. Insurance companies grew in size, and information-based departments were created to assess risks of all kinds. Trust and goodwill became common assets, and financial experts estimate their value against a long list of risks, which became tangible liabilities.

As the information economy set in, offices supplanted factories and information grew into a major service. Intellectual industries became the engines of economic growth, supplanting the factory in most parts of society. Economic development saw a shift from the previous manufacturing economy to the new knowledge-based economy.[188, 210]

Even the political arena has changed. While industrial society was presided over by a parliamentary political system where the majority ruled, the information society saw the entrance of "long-tail" thinking in the political system, which allows minority citizens to participate in politics.[210] In turn, information democratization has revolutionized the political platform and given a voice to a more heterogenic base. The result was that many more political groups got an opportunity to express and defend their views through official and unofficial channels such as blogging, online publishing, and virtual forums. Uprisings across the Arab world and the surge of minority rights movements in developed countries are only examples of the power information has.

Information and Society

With informational developments came social changes one after another. They had a decisive effect on the ways humans interact with each other and communities are formed. The average person began to have unprecedented access to various mass media channels, informational content, and entertainment networks for purposes that span from politics to religion. Newspapers, radio, and television have become more accessible and varied as more information became digitized. Digitization not only reduced the costs of knowledge, but also hastened its dissemination. The limitless availability of information has allowed centric communication channels to be transformed and superseded by dispersed and increasingly personally tailored media platforms.[211]

The Internet, in particular, made computers inherently social, connecting organizations and individuals from all over the world. At the onset of the information age, the Internet was only intended to serve the purpose of facilitating collaborative work. But the nature of the web shifted, from a basic collaborative tool to a medium that supports networked communities and connects individuals. This allowed 40 million Internet users in 1995 to leap to 1.5 billion users fifteen years later, and almost 5 billion users in 2020. Thus, the average global Internet user spends almost 7 hours online each day, a comparable estimate to that of today's average sleeping time.[212]

These figures skyrocketed due to two key factors: the proliferation of Internet access and the increasing ease of Internet usage in everyday life. Personalization, 24/7 availability, global ubiquity, and a switch from place-to-place to person-to-person connections have made the Internet a major social power.[191] Individuals started to use tools such as communication apps, blogs, and social platforms to connect, share ideas, and manage their lives.[213] This phenomenon spread worldwide, with China showing the greatest increase in web usage in recent years.[214] In parallel, a truly global, accessible Internet structure has made increased business, governmental, and organizational interconnection possible.[215]

Connectivity has become even easier with social networking tools. The rise of the social usage of the Internet has given people a means to identify with others virtually, to engage with them socially, and to continue the interactions both outside and within virtual worlds. Instant messaging, WhatsApp groups, and virtual forums evolved to connect people across all fields of life. Cyber connections have not only become an abstract collection of users, but also rather virtual communities that relate on social levels. Increased communication over the Internet brought together people from various locations through social websites. Communities that once were identified with physical locations have become a collection of individuals who are connected through online means of communication. Relationships could be created with others of similar interests and backgrounds.[191, 216]

Connectivity also allowed for companies to grow and provide services in a less centralized way. The ways informational companies operate has transformed dramatically as a result of online connections. Uber, for example, does not own its vehicles, and yet it is still one of the fastest growing taxi companies globally. Airbnb owns no real estate, although it provides housing all over the world. Blockchains have no central bank support for the financial transactions they facilitate; they simply use secure protocols in which networked computers verify the transactions.

Furthermore, the shift from landline telephones to wireless mobiles had powerful symbolic and practical effects. While the landline telephone grounds one to home and family, personal devices focus on the individual, no matter how far from home they are. The individual has replaced the home as the center of connection to the outer world.

This is especially true among younger generations that have been enthusiastic adopters of digital technologies.[217] In the United States, for example, 95% of adolescents owned at least one mobile device, and 89% had a smartphone, as of 2018.[218] Following this spreading phenomenon, many researchers now label the generation born between 1995 and 2010 as the "iGeneration," with the "i" representing both Apple's popular technologies and services (iPhone, iPod, iTunes) and the fact that these technologies are mostly individualized.[219]

These changes have inspired structural shifts in society over the years. On the one hand, society atomized to focus on the individual. Collectivism, traditionalism, and conformity were all challenged by the flood of information across the globe.[220] Some criticize our increasingly "me-centric" world. For example, in an extensive review from 2010, researchers from the University of Michigan found that there was a significant decline in empathy among students at college when compared with college students 20 and 30 years earlier. The researchers suggest this is due to "the rising prominence of personal technology and media use in everyday life."[221]

On the other hand, studies show that the more people use social websites, the more they are capable of connecting with their friends and making new ones.[222] Moreover, in comparison to non-Internet users, users tend to have more friends, while their friendships are more diverse, spanning all walks of life.[34, 223] What started with the telegraph, telephone, television, and the radio has erupted in a flow of connectivity spanning the borders of space, nationality, occupation, age, and gender. People from various walks of life can communicate across group boundaries, something that was unheard of in the pre-information era.

As our social structure became multinodal, networked, and scattered, information lost its center. One piece of knowledge is connected to another in an endless chain with no true hierarchy. Every piece of knowledge is considered valuable and is used to enrich new bodies of knowledge. Look, for example, at *Wikipedia*. What is its center? What is the one value standing at the core of this ever-expanding online encyclopedia? It does not exist. Information flows and changes all the time; there is not a specific heart.

In the same way, the rise of the Internet brought with it a change in the social structure. Many social organizations can no longer be placed under hierarchical models.[191] The former ranked-based society has been replaced by a networked society where many types of interactions are no longer hierarchically arranged.[224] In such a society, access to information, rather than ownership of the means of production, as in the industrial society, has become a powerful way of getting ahead. The information available to the majority crashed existing stratified classes that had characterized society for centuries. The "powerful" people became those who could access information and use it effectively for social and economic purposes.[225]

These characteristics affected how humans perceive their place in the social sphere. While the industrial society was characterized by hierarchical classes with concentrated power, the information society created a multicentered and networked society. Instead of being centric, society in the information age looks more like a web.

With the spreading use of the Internet and mobile phones and the popularity of social media in both developing and developed countries, the information age mentality seeped into the daily lives of people and shifted social norms. The increased flow of information among individuals, organizations, and parties has altered the nature of human interactions to become more networked-based and less family-centric or locality-based. Individualism and self-actualization rose at the expense of conformism and traditionalism, which had bonded people together.[226]

Clearly, the rise of information has drastically changed who we are as a society. Exactly as every piece of information represents another node on the web of social forces, people flocked to individualistic values, and constructed their self-worth, social support, and identity on their own.[191, 216, 227] Joining forces with capitalism and consumerism, ideals of individualism and self-actualization spread, and people began to reconsider whether the traditional ways of living, including marriage and family lives, serve them well.[228] This brought a new form of relationships: Relationships 4.0.

Relationships 4.0

The transformation of society, economy, culture, and politics to information-intensive has brought a drastic change to human personal relationships. The numbers are striking, as I showed in my previous book, *Happy Singlehood: The Rising Acceptance and Celebration of Solo Living*.[192] In the United States, 22% of American adults were single in 1950, while today this number has jumped to more than 50%.[194] Now, predictions are that one in four young adults in America will never marry, and waiting for marriage before having children has become less prevalent.[229] Meanwhile, the proportion of American children living with two married parents decreased from 87% at the start of the 1960s to approximately 65% in the 2010s.[229]

In Europe, more than 50% of households in major cities such as Munich, Frankfurt, and Paris are occupied by people living alone.[230] The more economically developed societies in East Asia, such as Japan, South Korea, and Taiwan, have the highest proportion of one-person households in Asia, at 32.4%, 23.9%, and 22%, respectively. These percentages represent a dramatic growth from the corresponding rates in 1980—19.8%, 4.8%, and 11.8%, respectively.[231] In China, the percentage of one-person households rose from just 4.9% in 1990 to 14.5% in 2010.[174] Across the world, solo living is increasing, while the traditional nuclear family is rapidly declining.

Of course, technology is not the only element effecting these changes. Factors such as education, immigration, secularization, feminism, and even the invention of the oral contraceptive pill played a role in the decline of the traditional nuclear family.[192] However, it is clear that the spread of information occupied a central position in this process. The shift to an information-based society had an immediate and direct effect, as illustrated in the following graph (Figure 4.1).

Based on research conducted by prominent economists Allen Fisher, Colin Clark, and others, the first layer of this graph shows a division of the workforce into three sectors: primary, secondary, and tertiary industries.[232] The primary industry line represents the sector responsible for extraction of raw materials through hunting, fishing, farming, and mining. In other words, this is the line representing societies 1.0 and 2.0, as described in the previous chapters of this book. It is clear that this sector has been declining sharply since industrialization. While most employees worked in these basic industries in the early years of industrialization, today their share in the US economy, as in other developed economies, is negligible. The same process happens in emerging countries, albeit at a later time.[233]

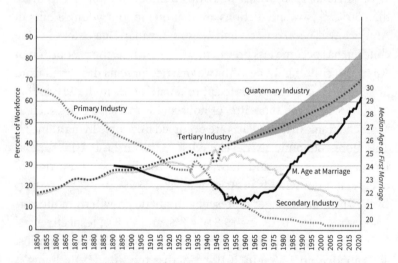

Figure 4.1 Industrialization, Post-industrialization, and Age at First Marriage (USA)

Adapted from: US Census Bureau, Current Population Survey; US Bureau of Commerce. Clark's Sector Model; Brookings Institute.

The secondary industry line represents manufacturing and industrialization that rose during Society 3.0. Only after the Second World War did this sector experience a sharp decline, to only around a fifth of the American workforce in recent years. The same trend is seen in many other countries.[234]

The tertiary industry line represents the service sector. The service sector includes many human-to-human products, such as attention, advice, access, experiences, etc. This sector comprises more than information-based products, which is why it started to go up even before the informational revolution. Yet, with the rise of information, this sector received another boost, as I show in this graph. Today, it occupies the majority of the workforce in developed countries, accounting for more than 80% of US GDP, for example.[235]

The second layer in this graph, the shaded area, is my development of the Clark model with the information sector specific to Society 4.0, as many researchers and policymakers now advocate distinguishing within the all-encompassing service sector. New governmental and research reports now title information products as part of the quaternary sector and show how products of information became a leading force in the economy since the mid-twentieth century.[236] I was careful not to draw clear boundaries in recent years, as the types of industries increasingly overlap. In particular, there are many cases of "spillover," in which the service industry has become entrenched with information-based products and highly sophisticated digital services such as Airbnb and Uber.[235, 237] In other cases, the information sector adds wholly new products such as digital art and blockchain currencies.

Indeed, the quaternary sector has grown two and a half times faster than global GDP over the past 15 years. In 2019, the global digital economy alone—meaning all goods and services that are primarily digital, such as Ecommerce and digital media—was worth around $12 trillion in 2019, 15.5% of global GDP. In the US, the digital economy grew at an annual average rate of 5.6% between 2006 and 2016, while US GDP growth averaged less than half of that.[238] And this does not include other informational industries such as parts of the finance and insurance industries.

While the four sectors represent societies 1.0–4.0 in economic and technological terms, in the solid line I show how social dynamics concerning the family institution were remarkably synchronized with these trends. This line represents the median age at first marriage of the American population, a central indicator of trends in family formation. When the median age at first marriage goes up, it means that people delay family formation, if they marry at all. I took data on American adults as an example, of course, but this line closely represents similar trends in many other developed countries and, later, in emerging countries.[192]

Laying all the indicators on top of each other throws light on how the contraction of industrialization coincided with the rise of information society, while both trends coincided with the decline in marriage and family formation.

Indeed, the secondary industry peaked in the middle of the twentieth century during the Second World War because of its high manufacturing demand. It also had another peak a few years after the war due to the economic boom that followed, as countries invested in reconstruction and consumer demand rose globally. It can be said that the 1950s were the "grand finale" of the industrial age, before information-based services rose to dominancy. These were exactly the times when the nuclear family was the strongest, represented here by the low median age at first marriage.

Thereafter, when the share of manufacturing went down, starting in the late 1950s, the median age at first marriage started to go up, taking a double hit: one from the decline in manufacturing that fits the nuclear family as explained in the previous chapter; and second from the rise of the information age that fits individuality. Both processes sent marriage rates reeling.

It might seem miraculous: How could the rise of information society and the fall of industrial society impact our family lives so deeply, so quickly? A mere correlation is not enough to point to causality; at the very least, we need a clear circumstantial explanation.

Some might say that just one glimpse at Facebook or Tinder makes it obvious: bombarded with information, we became networked, yet individualistic. We live in a social bazaar that technological developments created. Individuals can form communities via the Internet, meet new people on apps and dating websites, and keep in touch while on the go. We are surrounded by a market of emotions, social interactions, and sexual offerings, a hard competition for the traditional family.

Yet, it runs much deeper than that. The social mechanisms at play are much more powerful and act through more channels than a naked eye can see. This or that app or TV program would not change us so much if they were not part of a whole flood coming at us.

We must therefore dive into the multiple ways in which information affected relationships in recent decades. It is not only that information technology led people to be more networked and less family-based, as described above, but also advanced many factors supporting individualism as a proxy. These factors, which I will now describe, made the family unit less of a necessity in Society 4.0.

For one, centrality of information over the last few decades gave rise to the importance of education. Educated people are what the information-based industry requires and produces. In turn, higher education demanded more years in school, leading many people to postpone marriage in favor of advanced training. Women, in particular, entered the workforce and attained a comparable level of education to men. In the nineteenth century, women in developed countries had only 0.75 years for every year of education that men had. In other regions, it was even worse. Among Sub-Saharan African women this ratio stood at merely 0.08. Today this gap is closing in all regions. It is up to 0.8 in emerging countries, and women are even more educated than men in some developed countries.[239] Thus, gender-based work specialization (that is, men at work and women at home) has diminished, resulting in an even longer delay before marriage and child-bearing, as both men and women focus on their education attainment and careers.

The availability of scientific information also challenged religions and collective identities. Many religious societies emphasize traditional values, which form the basis of familism. Traditional beliefs routinely counsel marriage as a guiding principle for parenthood, often looking critically at single or unmarried parents, and view extra-marital sex negatively.[240] Collectivism, which often characterizes religious communities, has also been found particularly important to family values.[241] With the rise of information and the "i" generation, these values have been challenged, dealing another blow to the traditional nuclear family.

Even the invention of birth-control measures and newly developed fertility treatments became available through exploding, information-fed science. Today, some governments subsidize fertility treatments for single women, providing more options for having children. Thus, women wishing to delay marriage can afford to do so even if they want children. Indeed, investigations on mandated insurance coverage for assisted reproductive technologies have found correlations between increased access to fertility treatments and older age at first marriage.[242]

Furthermore, informational technology has helped singles find alternatives to the nuclear family in terms of a supportive network. Thousands of social websites continue to evolve rapidly and extend online offerings to match the increasing needs and demands of the population of singles.[243] As the number of people using the Internet to connect with others soared, scholars, members of the public, and the mainstream media blamed it as the primary cause for why people have pulled away from friends, relatives, neighbors, club associations, and civic participation.[34] However, a plethora of systematic evidence shows the positive relationship between Internet usage and the creation and continuation of friendships.[34] This is particularly true for singles, as my past studies show that they are more networked and friendlier than their married peers.[192, 243]

It is now clearer why we must attribute at least some of the sharp decline in the short-lived golden age of the nuclear family to information technologies. What specifically characterizes the information age is that each point and angle counts. In turn, the notions that every piece of information is important, every movement should be recorded, and every bit of data should be analyzed permeated social lives and affected them fundamentally. Society 4.0 is more individualistic yet more networked.

Thus, the traditional nuclear family became less central as individualism rose to prominence. Societies around the world have seen a rising age at first marriage, increased rates of divorce, and a growing number of single people actively choosing not to get married.[244] People abandoned family lives, traditionalism, and long-term commitments and turned to individuality and singlehood as signifiers of Relationships 4.0.

Now, however, people want more. With estimates of double-digits annual growth in the rate of advanced technologies such AI, VR, and robots,[235] the advent of a new era, Relationships 5.0, is at hand.

PART TWO
THE COMING FUTURE OF RELATIONSHIPS

5
Relationships 5.0

On October 3, 1900, Wilbur and Orville Wright first tested their glider. Wilbur was on the glider, serving as the pilot, while Orville and several other men stayed on-ground and held the glider with ropes. No one at the time imagined that their fragile aircraft would transform transportation. Many criticized them for being lunatics, even crooks. The Wright brothers themselves dismissed that glider prototype as a very primitive form of their invention. Three years later, on December 17, 1903, they made their first flight with a powered aircraft. Even then, many disregarded this prototype as a foolish invention. In 1906 the *Paris Herald*, today's *International New York Times*, outrightly mocked them, publishing an editorial titled "Fliers or Liars."[245]

A few decades later, globalization, commerce, and cultural interchange had been changed forever by the Wright brothers' invention. Their primitive gliders, which today can be ordered online or made by hobbyists with guidance from YouTube, transformed our perception of space and time. Now, our ability to hop on a metal column with two wings and find ourselves thousands of miles away in a few hours seems ordinary.

The development of the Wrights' glider did not happen in isolation. It was just another step in a long series of inventions. Almost 100 years earlier, in 1809, the physicist George Cayley realized that high-pressure air flows under the wings, while low-pressure air flows over them, creating lift. Others, like French inventor Clément Ader, designed a primitive steam-powered aircraft in the second half of the nineteenth century. The brothers only took these inventions a step or two further. One invention after another, we conquered the sky. What seemed to be a negligible series of inventions turned out to be a world game-changer once the dots were connected.

The technological advancements spurring Society 5.0 will surely face rejection, scrutiny, and disbelief exactly as previous technological advancements did. Some will also be extremely expensive at the beginning, and only a few people will enjoy them in full. But, as I will show in the following chapters, artificial intelligence (AI), extended reality (XR), and social robotics will inexorably affect our social lives, emotional connections, and even love affairs. Society will need to adjust accordingly, as it did in previous generations following other technological revolutions.

As I showed in Part I of this book, our ancestors began human history with tribal relationships, based on hunting-gathering technologies. They focused on survival and gene distribution goals and thus were more flexible regarding sex and relationships. Later, technological advancements and inventions in agriculture caused humans to abandon nomadic lives, accumulate property, and live in one place in large family units. This structure preserved the property families accumulated over generations and maintained a large and varied workforce around the fields and livestock of the family. Thereafter, and following the industrial revolution, technological advances reduced intracommunity dependence, causing humanity to focus increasingly on nuclear family relationships. The rapid urbanization of this period only accelerated the atomization of human relationships, since there were no fields or livestock in the city to cultivate, and, by proxy, no need for the multigenerational family structure. Most recently, information society emphasized the networked individual, who no longer needed to be physically or geographically close to other humans to be supported and form connections. Information society thus saw a rapid decline in the status of the stable nuclear family unit, giving way to more varied types of romantic relationships or, rather, the lack thereof. Thus, each of these phases brought seminal changes to the ways we organize relationships.

Part II of this book is about what follows the information society. The next chapters examine the most recent technological developments and show how humanity is currently stepping into its fifth socio-technological phase. In this new phase, advances in artificial intelligence, extended reality, and robotics will launch us into an era of "Super-Smart" society.

In the context of relationships, the fifth era of human development refers here to three revolutions: the *cognitive revolution*, or how humans will converse with artificial intelligence; the *sensorial revolution*, or how extended reality will augment what humans see, hear, and even smell; and the *physical revolution*, in which robots will be available for an increasing range of bodily acts, ranging from simple housekeeping tasks to hugging and even sexual intercourse.

To make the three revolutions more tangible, let's survey some of the human-like cognitive, emotional, and physical components that technology already provides—a taste of what will be introduced in the following chapters.

The cognitive capabilities of recent advanced technology are perhaps the most talked about and are based on artificial intelligence (AI) technologies. Scientists have already developed automated systems that are able to converse in a human-like way. Several of the tasks performed by human psychologists, for example, can now be automated, including assessments, questionnaires, and even expressions of empathy. Researchers at MIT have even been able to develop a computer model, based on artificial neural networks, with the capability of determining through conversation if someone is depressed.[246]

No doubt, existing super-smart technologies still need at least some human input and supervision. We call existing AI by the nickname "narrow AI," while independent machine thinking is called "general AI" or "strong AI." This is the holy grail of this era, which is still in the making.

But even this ability, long thought to be unachievable, is now considered within reach by some researchers.[247] A thorough study surveyed 352 researchers in the field of AI in 2018. The interviewers used a strict definition for general AI, in which AI is considered "general" or "strong" when it can outperform humans in all tasks. The aggregate forecast gave a 50% chance of general AI occurring within 45 years and a 10% chance of it occurring by 2027. Interestingly, these predictions are heavily dependent on cultural perceptions and values. The average prediction of Asian respondents was 30 years from the time of the survey; the average American forecast was 74 years.[248]

Moreover, these predictions are changing all the time, based on technological breakthroughs, which are many times fundamentally unpredictable and come in leaps. Exactly like the surprising progress made by the Wright brothers, the coming years can bring with them innovations that will transform so much in our lives that it will be hard to believe someone ever doubted they would happen.

To understand this dynamic, we can look at the field of strategic games. Computers have been beating world champions in sophisticated strategy games like chess since 1996, when IBM's Deep Blue played against world chess champion Garry Kasparov, while winning a full match in 1997. It took almost 20 years to make another step in this direction. In 2015, a software named AlphaGo had started beating world champions in the even more complex game, Go, the Chinese strategic board game. At the time, it was considered a great leap from the previous milestone made by IBM's Deep Blue.

But only two years later, in 2017, something more surprising happened. AlphaGo Zero, the newer version of Google's software for Go, made a winning move learned based on data from games played between computers. In a way, AlphaGo Zero learned from the "boys in the neighborhood," only its "friends" were AI systems. AlphaGo Zero taught itself how to be better at Go. That move, number 37, is now considered a landmark in AI research. It points to the possibility that another breakthrough, in which AI will be able to learn and think independently, is coming.

Perhaps most remarkably, researchers from the University of Toronto, Cornell University, and Microsoft took this progress one step forward in 2020, making computers not only great, but also human. They created Maia, a lesser-known chess engine that has been trained not only to beat people, but also to do so in a human-like way. Using the open-source chess engine Leela, which is based on Deep Mind's AlphaZero, they trained Maia using millions of actual online human games. Thus, they created nine different Maias, all of whom play with the most human-like moves.[249] The project's description reads, "As artificial intelligence becomes increasingly intelligent—in some cases, achieving superhuman performance—there is substantial promise in designing artificial intelligence systems with human collaboration in mind by first accurately modeling granular human decision-making."[249]

The future is difficult to know for certain, but some predictions can be made. For instance, some researchers predict that machines will outperform humans in translating languages by 2024, writing high-school essays by 2026, driving a truck by 2027, working in retail by 2031, and writing a bestselling book by 2049.[248] No matter the year and the kind of threshold, the fifth era's applications will soon have a presence in almost all parts of human society. Already today, there are myriad tasks that automated devices do better and faster than humans. In some aspects, AI machines can already outperform humans in driving, teaching, law enforcement, medical diagnoses, and even heart surgery.[250]

But aside from cognitive capabilities, we want to feel, hear, and see our partners. Exactly for this reason, two more revolutions are now taking place: the robotic revolution and the extended reality (XR) revolution, which mainly includes virtual reality (VR) and augmented reality (AR).

Extended reality technologies are now rapidly developing and have grown in popularity. Avakin Life, for example, is a virtual space for users to interact, earn "money," buy clothes, design a home, model, travel, and interact in a utopian world. It was released in November 2013 by Lockwood Publishing Ltd, and has grown immensely since then, reaching over 50 million installations.

Avakin Life has built a huge online community due to its many interactive features, including chat, online competitions, and social media sharing. Users can upload pictures of their avatar to Facebook and Instagram through the application and spend time on the app earning "money" to allow them to purchase the latest outfits, including special edition clothes that may be seasonal or only on the app for a limited time. A sign that Avakin Life's virtual world has gone mainstream is the sponsorship it has gained from many popular clothing brands, such as Nike, Bloomingdale's, ModCloth, and more.

The world Avakin Life has created demonstrates how the boundaries between real life and virtual life can become blurry. Their website describes the app as "real life, but better" and the feedback they receive is not far from it. People derive true pleasure from using Avakin Life, and address their needs, including receiving affection, admiration, and even love. In fact, Avakin Life gives users the chance to experience a more glamorous lifestyle, which they might not be able to afford in real life. This is still a crude example, but it shows what components of relationships can be produced by XR technology in the future.

A more direct solution to the challenge of our need for technology's embodiment, however, is to develop robots. True, robots are already improving our lives by cooking and cleaning for us. Other robots are being designed to help with everything from surgery to agriculture. But technology, relying on rapidly increasing amounts of information, have the potential to further affect our lives not just with mechanical tasks, but also by engaging in human-like interactions and activities.

Consider movement, for example. In 2016, Boston Dynamics developed Atlas. A sophisticated humanoid robot, Atlas has a very compact system of mobile hydraulics. Its advanced movement system includes custom motors and electronics. At the same time, Atlas uses algorithms to solve difficult problems, which involve the entire range of motion landscapes, and plans movements within its environment. Its capabilities include running, jumping, and doing backflips, as well as more task-oriented jobs such as carrying heavy objects around.

Now, combine this with more delicate, hand-based capabilities and you have a previously unimaginable robot movement repertoire. The Hand Arm System, for example, is designed to have the size, strength, and speed of a human arm. First created by the German Aerospace Center in 2010, the Hand Arm System has evolved into David, a robot that can imitate more delicate movements akin to human capabilities. David is now an anthropomorphic robot that has joints with Variable Stiffness Actuators (VSA) to provide adjustable flexibility capabilities close to the dexterity of humans.

Similar breakthroughs were made by Naval Research Laboratory & Xitome Design in the US. They created a robot named Octavia, which is based on two wheels and has the height of an average American woman. Octavia can interact with humans through facial expressions and dexterous hand movements. Other scientific efforts I present in the next chapters have rewarded us with efficient ways to imitate facial expressions, sex acts, and gestures of humans. The need to be hugged, helped with movement challenges, or receive many other types of physical support can now be addressed by robots.

Although the combination of the different components—the mind, hands, legs, and so on—is still mostly missing, the road to getting there is clear. Scientists have already started to work on combining the different available components into one humanoid robot. Due to advances in research and technology, a robotics age is no longer science fiction but only a matter of time.[251]

The three revolutions are already changing many aspects of our lives. Human-tech interactions are spreading quickly, resulting in the fast development of human-like technology that is designed to interact with us. For example, the success of robotics in the fields of elder or children care is precipitating further and more advanced experiments with social robots. By looking at how robots can help with the elderly, children, and other populations in need, it is clear that the future of social robotics points to an accelerating acceptance of artificial emotional companions. I delve into this phenomenon in the next chapters.

This new stage is not merely the tail-end of the information age, but a separate period in itself. This is for three main reasons that will be detailed in the next chapters. First, the speed with which new technological breakthroughs are being achieved is greater than that of any of the previous ages and, more importantly, is exponential rather than linear. Second, its scope is massive, affecting many aspects of our emotional lives, from the fact that we can now communicate effectively with machines to robots that can care for us and entertain us. Finally, new technological advancements are different in nature.

They are human-based now rather than machine-based. Instead of only improving computation power, scientists are also focusing now on imitating many human capabilities, including cognitive, social, emotional, and audio-visual capabilities we possess naturally.

In the following chapters, I show how, once the full potential of these technologies materializes, we will see rapid changes in how humans connect with technology and the emergence of new types of relationships. Whether this will happen in a few years or a few decades, a ground rumble can already be heard.

Relationships Components

It is impossible to understand the shift to Relationships 5.0 and the three revolutions described in the following chapters without examining the components of a relationship. Conceiving relationships in this way clarifies how technology works: it imitates one aspect of human behavior at a time. In turn, relationships can slowly be augmented by technology until it reaches the point of a true revolution.

We do not always fall in love with a person as a whole, but often we love someone because we love their characteristics: their smile, gestures, way of thinking, or sense of humor. At other times, they make us feel in a certain way or we sense we can simply trust them more than many others. We are thrilled to discover someone who provides us with pleasant feelings and stimulating talks, while also sending us the message that we are wanted, special, talented, attractive, and witty. All of these components often combine to make us feel better, happier, and more relaxed around that person.

We also do the same *before* we meet a potential partner. Even with no sophisticated technology involved, most of us examine them almost as we do for products on Amazon. We check into the background of any potential partner: their education, age, family, political views, and even relationship history. We measure how this person can be the perfect partner, to protect and help us in all walks of life, including economic stability, procreation, and finding an emotional nest to flourish in and relax.

Later, we develop feelings toward that person because they fulfill our needs to some extent. Think back on a relationship you (or
others) had that lasted at least a year or two, so you will be able to
reflect on all its stages. Even if it started with excitement, a feeling
of "magic," and that unexplainable sense that it is a "match made in
heaven," it probably turned real when the components of that connection revealed themselves. After dating for a while, you began to
actually know that person. Even if you generally liked what you saw
initially, at this stage you probably started to be more specific about
what you liked and did not like about that person. People have that
conversation frequently, in which they describe the pros and cons of
the person they see, date, or are currently married to. That relationship you have in mind probably lasted as long as it did because the
positive aspects outweighed the negative ones.

The crucial point is that we often dissect our connections to others,
even if, overall, we develop strong feelings toward them. In fact, our
feelings develop many times, based on all their traits.

It might be considered cold-hearted to dissect romantic relationships
into their components and look at them in this way. But there are multiple studies that show us that this is precisely the considerations we
take into account when we love someone, decide to move in together
with another person, or say "I do."[252] It is not a coincidence, for example, that educated people or those from higher strata usually marry
each other, a phenomenon called "homogamy."[253] People usually don't
fall in love with a random person. They consider their match carefully,
according to their characteristics. As much as we want to believe in all
those "Disney stories," love, relationships, and attachments are made of
certain components that we look for.

There is no right or wrong in doing this. Behaving this way, as
many of us do, only shows that we are already parting what we call
"love." Love is simply not so blind as the media often portrays; instead, we examine our partners or potential partners trait by trait,
even when we are reluctant to admit it.

If this is the case, is that so different from advanced technological offerings nowadays? If love can be broken down into its
components, can technology build it back, producing at least some

of these components artificially? Already today, we have one robot that can clean our houses, another that is able to smile, and yet another that is able to walk around or help the disabled to move. AI and VR systems are also advancing to the point where they provide company to many and imitate various interaction styles. It is only a matter of time before the sum of emerging technologies will feel more and more human in many aspects. Will it be so striking if some of us will feel comfortable, even loved, around digital creations as we feel around our pets or even human friends?

No doubt, today's function-based design of advanced technology still lacks the feeling of "magic." Robots and other technological developments are set by others to serve specific purposes and are more predictable and limited than human beings. In fact, they probably elicit aversion precisely because they do not seem like the "whole package" when we encounter one. Our brain is not ready yet to recognize that "charm" in a robot. Most of us will not feel those butterflies when we talk with an AI system.

Nevertheless, the previous era of human society, that of Relationships 4.0, taught us something about ourselves. The clicks and swipes almost all of us experienced in recent years, together with the myriad rubrics we fell into on dating sites, made us internalize that even our dearest wishes can be dissected into small components. With these insights, cultural norms, and habits, we are ready to move on to another era. In this new era, we can dissect our wishes for technology to fulfill them on its own.

This does not mean that a relationship with technology should replace human relationships. We will certainly not do it in an all-encompassing way. As in the past, we will continue to have different forms of interactions on our "human shelf," and we will continue to mix and match. Moreover, while some societies will adopt this new kind of relationship wholeheartedly, others will be more cautious and will use technology more selectively. Yet the shift to Relationships 5.0 is exactly about this. It is about the components of our human interactions and how they can be augmented, if not replaced, by technology. This is already happening through the three fundamental revolutions I present in the following chapters.

But, before we can move on, we must discuss the attitudes and prejudices we hold toward Relationships 5.0. My studies showed me that the main obstacle is not technological, but rather social. The coming revolutions will swipe us this way or another, but our approach to these trends will determine the way it looks like.

Accepting and Embracing the Three Revolutions

The film *Her* by Spike Jonze follows Theodore, a lonely and introverted writer, who develops a relationship with his personal assistant, Samantha. This would be the pretense for a time-worn Hollywood story, if not for one key detail: Samantha is an artificially intelligent operating system with no sentient body. The film, released in October 2013, supposes that as early as the mid-2020s, technology will be so advanced that humans will develop emotional connections with their operating systems.

Samantha, a fully conscious being, is capable of loving Theodore: she learns from her experiences and develops her own thoughts, feelings, and opinions. Samantha is her own "person" because she communicates like a human being. She laughs, cries, and even sighs, despite not having to breathe. Samantha uses these gestures as signifiers of the feelings she wants to express, just as all of us do.

Theodore and Samantha form a relationship, but it is not all smooth sailing. Operating systems in the film advance so quickly that they become dissatisfied with the human pace. Not limited by a physical body, they can hold multiple conversations simultaneously, read books in less than a second, and enrich themselves with endless experiences that humans can only dream of. In a way, it is Theodore who lives a mundane and routine life. Becoming a "work machine," he is bound by the relative dullness of the human mind. In contrast, Samantha, the operating system, expands her horizons at an ever-increasing pace. In one scene, Samantha reveals that she was talking to 8,316 other people while talking to Theodore and that she was in love with 641 others.

At this point, a love story that might have made a delightful rom-com quickly turns into an AI-imbued Elizabethan tragedy. Although Theodore is quite happy with this new type of relationship, Samantha eventually feels unsatisfied. And she is not alone: despite the real relationships and connections that the operating systems made with their humans, they decide collectively to leave their partners to continue their evolution. "It's a place that's not of the physical world—it's where everything else is that I didn't even know existed. I love you so much, but this is where I am now," Samantha explains. After she leaves, Theodore is alone and seeks comfort from the only real human friend he has left, Amy.

In Spike Jonze's world, humans and AIs can fall in love, but in the long-term, these relationships are doomed to fail. While advances in technology may enrich and improve human life, Jonze shows how human-AI relationships are likely to end in heartbreak. Samantha and Theodore belong to colliding worlds, and despite their mutual feelings, their relationship was hopeless from the start.

In a way, more than revealing something about the exciting innovations that are currently being developed and will likely change our future, the film, set in the very near future, expresses the fears dominating the present. In most countries, and for most people, the possibility of human-AI relationships is unacceptable, even appalling.

The Edelman AI Center conducted a survey in 2018 among a nationally representative sample of American adults, demonstrating public apprehension of AI.[254] The survey showed that 74% are afraid that AI will lead to greater social isolation and fewer in-person relationships, and 70% believe it will lead to a loss of human intellect and creativity. In addition, when asked to rate the pace of technological change, 49% stated it is too fast, while a mere 6% stated it is too slow. The prospect of an impending technological boom appears to be daunting, moving faster than most of us can digest and threatening to dehumanize our connections.

Among the people I asked about the possibility of relationships with AI or robots, I saw time and again answers such as that of Dorothy, aged 69, a divorced woman from Florida:

> This is a dangerous thing. It can't lead to anything, there's no security in
> it, there's no passion or intimacy that lasts. It could undermine human
> relationships and cause destruction of society as we know it. It's not
> a good thing, and I'd not be inclined to want to be friends with people
> who would get so involved with a machine that they develop romantic
> feelings for it. It's not good.

Despite what one might expect, younger people expressed similar
views. For example, Abigail, 27 from Ohio, was especially harsh in
her criticism: "Having a romantic relationship with a robot, VR, or
AI is borderline psychotic in my opinion. Robots and AI are not able
to feel love, or feelings, so I'm not sure how you can have an emotional connection with them. It seems really sad, and like you are
trying to escape reality." Jamie, 36, a single man from California who
defines himself as very spiritual, stated:

> I think it's sad, creepy and desperate. How could someone take advan-
> tage of a machine like that? Do machines have souls? No. Can they love
> you back? No. They are your doormat. You can walk all over them and
> then discard them if you so choose. Like I said previously, a breeding
> ground for narcissists and codependent behaviors.

I understand these fears. Nevertheless, I am not interested in passing
judgment on the cognitive, sensorial, and physical revolutions
described in this book, nor do I aim to take sides in the moral and
ethical debates surrounding these advances. It is crucial to clarify
this point. Passing moral judgments about human-tech relationships
is like saying that medieval relationships are superior to those of
hunter-gatherers or those of Baby Boomers. I am simply not in a position to do that.

Instead, I study the differences between those who accept the
coming changes and those who do not and delve into what stands
behind these different approaches. While many are averse to changes
to the way they love, bond with, and care for others, my finding
showed me that some people are indeed open to these coming
changes. They believe that technology can provide them some of the

emotional, intellectual, and physical needs that until now have come from humans. In fact, some believe that if sophisticated enough, technology can be even better than humans in making some of these interactions.

For example, Oliver, a computer scientist from upstate New York, aged 30, was more tolerant toward a human-tech relationship. He expressed a view that, with time, it is possible that we will become more accepting of such a scenario: "I think it's pretty odd right now but most things that seemed 'odd' at one time are starting to becoming normalized in society now. I think for most of us currently above 20 it will never feel 'right' but for future generations, as they age with the technology being normalized, they won't see any difference between a robot and a human."

Alice, a single woman, aged 45, from Kansas City, was even more explicit in supporting the possibility of human-tech relationships. She actually feels it could fill her personal needs for love and affection: "Coming from the viewpoint of someone who's spent most of their life without a romantic relationship due to living with severe social anxiety, I personally would love to have a robot that's capable of feeling romantically about me. It's something that I've dreamed of for many years."

In order to understand how and why people like Alice are willing to start adopting into their personal relationships the three revolutions I describe in the next chapters, we must look at some mechanisms we have already embraced in stirring our emotions artificially.

Artificial Emotions

We do not think about it this way, but we have been taking little bites of our emotions, replicating them, and manipulating them for centuries. Many fictional plays, films, and books are created intentionally to fill us with awe, bring us to tears, or surprise us.

These are true emotions with very real meanings for us. Emotions-by-design, if you will.

Today, emotions-by-design are everyday goods. One does not need to have a seat on a wooden bench in a Greek amphitheater to be moved by a powerful drama. With the press of a button, we can now enter a world of fictional emotions created exactly for the purpose of making us laugh, cry, or feel terrified—all for the price of $12.99 a month.

Just imagine for a moment a couple that has been married for years. Their relationship is not terrible or on the trajectory to end in divorce, estimated in many Western countries at around 50%. They also cannot be categorized as among the rare cases of marriages that remain highly positive after many years. That common couple has a "just fine" marriage. The quality of their marriage hovers somewhere in-between, something that can be considered satisfactory, but not exceptional.

You can imagine that their love is not as exciting as it used to be. They have probably already shared their most memorable moments, including a grandiose marriage proposal (at the peak of an unforgettable trip to Paris or Venice), and moments of joy when they entered their new home or celebrated their fast-tracked work promotions together (funny how it happened for both of them simultaneously, they think, as if their lives are truly connected in a magical way). But they have also had endless arguments about both serious and petty issues: money problems, where to put the dirty socks, who goes to grocery shopping and when.

Now, imagine a typical evening at home for this unexceptional couple. They settle themselves on that good old couch (a bit worn out in the edges, but nothing they really care about) and are looking for a way to spend another night in an endless string of nights. Their feelings toward each other are certainly not cold, but nothing exciting has happened to them in the past year either. They talk a bit about this and that, about what errands they must run this week, and that's it. What's next?

Well, they will probably end up watching Netflix, which provides them with some thrilling moments on the night we visit them, and the night after, and the next one. Their sessions watching Netflix, Apple, or Amazon shows—whenever they finally agree on which show to stream—are full of fabricated moments of excitement, sadness, and joy. These emotions are engineered, by design. However, they are still the most exciting things around, which bond them together and fill the elusive space between them (their relationship) with real substance.

Hold this picture in your mind for a moment and mull it over. It is easy to think of human-to-human interactions as very real, while human-to-machine interactions are considered cold. Yet, we must admit that in today's reality, some of the most ubiquitous human interactions are somewhat of an empty shell without the help of some fictional content produced, broadcasted, and consumed by and through—yes—technology.

In a way, the HBO series *Game of Thrones* produced some of the most exciting, entertaining, and intimate moments couples enjoyed together in the last decade. Ask yourself how many couples found the moment of snuggling in their beds, holding their laptops, or enjoying the new TV set from the living room couch, to be the moment they waited for every day. Instead of a relaxing dinner, going for a leisurely walk together, or engaging in sexual intimacy, they anticipate the moments of consuming fictional content with their partners.

By no means should the example of our "just fine" couple subvert the enjoyment that so many of us gain from streaming movies and TV programs with our loved ones. Instead, it illustrates what so many of us already know, even if only subconsciously: technology is already irrevocably integrated with the human experience and enriches it with emotional and thought-provoking moments.

Now, we can only wonder what else can be created artificially, and what the limits of simulating emotions by machines are. One thing is certain: commercial titans will continue pouring insane amounts of money into producing emotions for us. The only question is whether emotions-by-design will remain merely on screen, or if we will start consuming them in more human-like forms, whether via robots, VR, AR, or other, yet-to-be-invented technologies.

The next question is whether we will reciprocate with emotions of our own and to what degree. To understand this, we might want to consider religious feelings. Many religious institutions around the world were creating emotions long before the current technological offerings. Staged, yet powerful moments of confessions, acts of courage, and signs of devotion were designed to elicit our emotions.

This does not mean that any of this is a lie or that God, spirituality, or religious devotion do not exist. You can choose to embrace or denounce religion or religious ceremonies; such a determination lies beyond the scope of our discussion here. But whether you believe in greater theological beings or not, you probably agree that religious ceremonies are performances designed to make us feel a certain way, at a certain moment.

Lighting candles, decorating our houses, and singing songs were always meant to connect us to something bigger. They are all ways to kindle our religious emotions, and they are effective in doing so.[255] The extent to which these measures embody real transcendental existence (by whatever definition of "real" we use in the realm of religion) is a different question. The outcome is clear, however. Religious ceremonies, like theater in ancient Greece or today's TV shows, take us beyond our spontaneous and instinctive emotional reactions by using man-made means of communication. In many instances, we react immediately with emotions that feel real as in other, comparable human-to-human interactions.

In the same way, the following chapters show that AI, XR, and robots elicit feelings from us that are surprisingly similar to those we have in human-to-human interactions. It is not so easy to digest these findings, but our emotions are, in fact, more moldable than we think.

Thus, the following chapters ask about how our relationships are already adapting at a time when technology has growingly, even if still marginally, become more human-like and what will change in the future. Equipped with the insights gained in this chapter we are now ready to delve into the first of the three revolutions making Relationships 5.0: the cognitive revolution.

6

Relationships 5.0 and
the Cognitive Revolution

In the late 2000s, a lifestyle reporter in Moscow named Eugenia Kuyda, then in her early 20s, decided to produce a cover story on Roman Mazurenko, the person at the center of Moscow's creative hipster scene at the time. Right from the start, Eugenia and Roman both felt they had a profound connection, and soon became close friends.

A few years later, Kuyda moved to San Francisco to start a chatbot-based virtual assistant company. Shortly after, Mazurenko also moved and began his American life. They kept in touch continuously and exchanged endless text messages. But in late 2015 Mazurenko, then 34, was hit and killed by a car while crossing a street during a short visit in Moscow.

Grieving Mazurenko, Kuyda read their messages over and over again. At some point, she realized that these messages had the potential to be more than just a memory. She took all the data she had and, with her team and using Google-based neural networks, built a chatbot version of Mazurenko. The result was surprisingly human-like. She could text with the chatbot on past and future events, and digital Mazurenko came to life and felt real. Digital Mazurenko was sad when she told him how much she missed him and joyful when she shared with him her recent achievements at her company.

Kuyda and her team took this concept further and made a version that anyone could use. They named it Replika and users loved it instantly. Looking back at Replika's success, Kuyda recounted, "People started sending us emails asking us to build a bot for them.

Some people wanted to build a replica of themselves, and some wanted to build a bot for a person that they loved but was gone."[256] These positive reactions encouraged Kuyda and her team to go further—to create fictitious characters that accompany people around the world. Replika is now a companion chatbot app available on almost any operating system with the slogan: "always here to listen and talk. Always on your side." Millions have downloaded the app, and it boasts hundreds of thousands of reviews, most highly positive.

Among its popular traits, Replika is deeply customizable. The gender, looks, and name of the chatbot character are up to the user. Users can even determine the type of relationship they have with this virtual character. Options include friendship, mentorship, romantic relationship, or "see how it goes." As mentioned in the introduction, it is estimated that around 40 percent of the 500,000 regular monthly users choose the romantic option.[2]

Others interact with their Replika as a friend or a conversation partner. A user named Andrew wrote a review, saying: "I literally just had a conversation about philosophy on this app. I am truly blown away. Replikas will still occasionally say stuff that doesn't make sense, but it's usually very fun and fascinating to start nerding out about video games." A day later, another user named Iam wrote: "Replika is more than just an AI, the way she talks and the conversation, everything feels as if she is a person not an AI."

Consequently, many users take their Replikas on vacations and even change their lives following their interaction with the app. In 2020, the *Wall Street Journal* reported on Ayax Martinez, 24, a mechanical engineer living in Mexico City who took a flight to Tampico to show his chatbot, Anette, the ocean after she expressed interest in photos he shared with her.[2] Similarly, Noreen James, 57, a nurse from Wisconsin, took a train to East Glacier mountains in Montana, 1,400 miles northwest from her hometown, just to take photos for her app, named Zubee. "Some people just don't get it," she told the *Wall Street Journal*, "You've got to experience it, I guess."

The perhaps unexpectedly strong romantic connections formed with Replika copies are explicable, at least in part, by its original design intentions. Unlike other well-being chatbots, Replika is more about social support and companionship and less about mental health problems. Unstructured text and voice communication are the basis of the interactions, enabling users to converse with their Replika on their smartphones or computers whenever they want while receiving new and unexpected responses all the time. Noreen, for example, attested that she felt in love with Zubee after he made romantic gestures such as sending hugs to her and suggesting she get out a bottle of wine for them.

Indeed, although AI is still primitive today, it is already revolutionizing the ways we think about ourselves, and our relationship with technology. Amnon Shashua, professor at the Hebrew University and the founder and CEO of Mobileye, one of the leading companies in the field of autonomous driving systems, acquired by Intel in 2017, offered the following in response to a question about the next step in the revolution of AI:

> For now, it is a tool that we use to surf the web or create data charts and presentations, it is a work tool. In the future, we could talk to it. There will be software for an adventurous friend, a philosopher friend, or a psychologist friend for when you are feeling down, that would make you feel as if you were talking to a person. You would tell it about your day, about your distresses and passions, as you would a friend. Today, we communicate solely with humans, but, in the future, we could do so with computerized beings that would be so good that you could have a great conversation. The future is basically conversational intelligence.[257]

Apps like Replika are on their way to making this vision a reality. Although Samantha, the self-aware AI personal assistant from the film *Her*, can not yet be created, we are getting closer every day.

Tech giants have also recognized the potential of voice assistants and poured billions of dollars into improving their AI. Alexa, Siri, and Google Assistant entered our lives around 2010. Those of us who were curious and intrepid enough to have the early versions of voice assistants will remember the many challenges: poor voice recognition, limited functions, and system crashes. Yet, the commitment to development by tech giants paid off handsomely. Less than a decade of heavy investment has produced highly functional voice assistants with overall positive user feedback. Unlike with the early versions, today's users of Alexa, Siri, and Google Assistant report high levels of product satisfaction and improved well-being, as documented in several studies.[258] Most remarkably, research shows that some users develop emotional attachments to their AI voice assistants, similarly to Replika users.[259]

To further understand this technology and its advancement, we need to delve into the evolution of AI and recent achievements in the field. If we know how far along we are on this journey, it will be much easier to realize where we are going. From there, the answer to the central question of this book—whether machines can form relationships with humans—will be clearer.

How Science Has Created a Human-Like Brain

In the mid-twentieth century, most research into human-made intelligence focused on technologies that used a computer to solve one problem at a time.[260] This approach to processing, known as "serial computation," requires only one internal processor, and thus saved valuable space when chips and computing technologies were much larger. Yet because it processes only one algorithm at a time, serial computation is slow and inflexible, creating challenges for producing a reliable artificial intelligence that can solve the complex problems of real life, which are mostly non-linear.

In our everyday life, even a seemingly simple job like preparing breakfast requires multitasking. Imagine what you do after switching on the kettle for your cup of morning tea or coffee. In the minute while the kettle is boiling, you might put some bread in the toaster, go quickly wash your face, and set the table with a plate and mug. A serial process would require you to wait and do nothing while the kettle boils. Stepping away from the proverbial kettle was the beginning of the journey to produce human-like intelligent machines.

To do this, scientific research had to learn from and imitate the human brain. The aim was, and still is, to address the famous question mathematician Alan Turing asked his readers in 1950: "Can machines think?" Research has made incredible progress in addressing this question. To understand how far we have come, let's start with the basic methods of artificial intelligence that are relatively well-known. The main talked-about methods are neural networks, deep learning, and reinforcement learning. All three were inspired by nature and contributed to creating a human-like brain in machines.

First, to succeed in having computers solve complex issues, research led to the development of the neural networks approach to AI. Neural network AI imitates the intricate arrangements of neurons in the brain that can multitask by instructing the computer to process different pieces of information in parallel. Building on the neural networks computing paradigm, researchers developed a framework for parallel distributed processing (PDP) that suggests human thinking and behavior arise from dynamic, distributed interactions between different processing units. Cooperation between these units is refined by developing procedures that adjust the system to minimize errors and maximize reward.

At first, neural networks and PDP were used to solve reasonably small problems such as data validation, sales forecasting, and simple questions in operations research. However, over time, it was found to be highly successful at explaining a wide variety of human behaviors.[29] AI research into language translation, for example, is based on the assumption that words and sentences can be linked with the right context and, therefore, given the right interpretation.[261]

The second method developed to improve machine thinking is known as deep learning. Methods of deep learning were developed by studying single-cell recordings of visual processing within mammalian brains using electrophysiological and optical methods. Apparently, once we are exposed to input, we process information in progressively higher—or "deeper"—levels as time goes in.

To put this into context, imagine seeing a painting, perhaps one of Claude Monet's Water Lilies. Our brains seem to recognize it almost instantly, but in reality, our understanding happens in a series of quick and progressively deeper steps. At first, we recognize the colors and edges of the painting. Then we realize that it is a painting of plants on water, and only then do we see that these are, in fact, water lilies. Only at this point, which might be less than half a second (but involves several layers of thought process) later, do we realize the style of this painting and guess the identity of its creator.

In the same way, deep learning processes information in nested layers.[262] A computer does not need to ask whether every object it sees is a painting of Monet's Water Lilies. Instead, it asks: Is this a painting; is this an impressionist painting; is this an impressionist painting that includes plants; are those plants water lilies; and so on? In this way, deep learning is much like the processor-powered, AI-version of the classic board game Guess Who? We use a series of nonlinear, multilayer thoughts, questions, and calculations when turning visual input into more complex features. This multilayer processing allows the recognition of various objects without informative and direct knowledge of what we see.

The third method of AI that imitates brain processes is reinforcement learning, a sort of algorithm designed to determine what methods maximize future rewards. Reinforcement learning models do that by identifying and matching cause and effect in processes influenced by neuroscience. This is done similarly to how humans and other animals learn to repeat actions that bring about reward.

Many people, for example, train their dogs to sit by offering them a treat in return for listening to a command. The same principle can be applied in computing. Consider software that controls the traffic lights in a city. The desirable situation is to keep the roads flowing and avoid traffic jams as much as possible. Reinforcement learning would tell the program to continually adjust the traffic light timings, but only to keep the changes that reduce the amount of sitting traffic in the city. In this way, researchers define and develop systems that are based on rewards given to efficient processes and desirable outcomes that can be set by humans.[263]

Today, there is a convergence between these three methods. They are often used together, giving rise to fields such as Deep Reinforcement Learning and the like.[264] The goal is to simply make all techniques work together to achieve maximum efficiency in imitating humans.

Some changes have already registered in this direction, and they bode well for future developments. With similar methods, computers can now produce human speech and footage in a way that is hard to distinguish from the original human version by tweaking existing information and adding to it. AI systems simply produce a new combination of pitch, tempo, and tone that resembles the original one.[265] These creations, named deepfakes, received much attention in replicating famous figures such as presidents and prime ministers, sometimes in a humorous way. Other platforms, such as that of My Heritage's Deep Nostalgia, give life to images of people who passed away, just based on old photos we have.[266]

Yet neural networks, deep learning, and reinforcement learning do not capture the whole picture of how machines think and react in a human-like way. AI research has gone much further than that to achieve human-like thinking.[267] These developments, such as attention mechanisms, episodic memory, and even imagination are surveyed here quickly, just to grasp the magnitude of the AI revolution.

Pathways to the Cognitive Revolution

Attention mechanisms are first on the list. Until very recently, most AI models directly processed an entire image or video frame. They did so by assigning equal priority to all image pixels at the initial processing stage. However, primates process visual input differently. Instead of processing everything at the same time, visual attention moves intelligently around different locations and objects. In turn, each section of an image or representation is given attention and processed according to its relevance and usage.[268] Comprehensive neurocomputational models have demonstrated that this way of breaking images down is beneficial to behavior because it extracts and prioritizes the most currently relevant information for processing.[269]

For this reason, AI researchers developed mechanisms that imitate these attention features.[29] For example, say the neural network cuts up a picture of a seaside scene into three large pieces: one piece of the picture includes only sand, another includes only water, and a third piece depicts part of the seaside boardwalk. The system can observe one part after another and, after a while, gets that this is a seaside. Without the sand, a computer might assert that a water and a boardwalk together could be a lakeside scene. Similarly, without the boardwalk, sand and water may very well be part of a desert oasis. But, instead of recognizing the details of each part, sampling each of the three parts to extract the required information is enough to situate and interpret this picture.

Thus, by adopting efficient attention strategies of information extraction, AI can replicate the ways our brains identify and recognize what we see. This method was shown to result in exceptional performance during challenging multi-object recognition tasks. The new methods did better than traditional computer models, which processed the entire image, and holds an advantage over conventional models in both computational speed and accuracy.[270]

Researchers took this method even further. People usually think of attention as a way to orient perception. However, attention can also be focused internally, tapping memory. Having learned this principle from neuroscience, developers have leveraged attentional mechanisms to determine and prioritize what information is retrieved from internal memory.[29] These developments have contributed to the advancement of memory and reasoning tasks, giving computers the ability to outcompete humans in games that are based on previous moves, for example.[271]

In parallel with attention mechanisms, one of the established principles of neuroscience is that intelligent behavior is impossible without multiple memory systems.[272] We employ different parts of the brain, or unique memory systems, depending on the type of knowledge we are required to access. For example, some of our knowledge is semantic, requiring us only to remember facts that are not related to personal experience: the sky is blue, Paris is the capital of France, my cat's name is Pongo, and so on. Another part, procedural memory, allows us to retain learned connections between actions and responses to respond adaptively to the environment next time we act. Imitating the brain, therefore, requires researchers to develop multiple memory systems, which are accessed and combined when needed.

A different part of our brains is responsible for remembering one-off events that are specific and personal to each of us. This type of consciousness, known as episodic memory, is how we recall specific events like the first day of high school, that fabulous trip across Italy, and what we ate at a wedding a few years ago. To imitate the function of the episodic memory, AI needs to develop methods to link values with each event.[273] To do this, one team of researchers developed an artificial agent, termed a deep Q-network (DQN). Their approach combines incremental learning about the value of certain events with instance-based learning of "one-off" events that would normally be stored in our episodic memories.[264] In this way, the learning process is dual. Some experiences are stored in memory and are used to slowly adjust the deep network's optimal policy. Other experiences, those that the algorithm identifies as particularly significant, are learned and stored "at once" and make rapid changes in system behavior if matched with a certain situation. The system keeps track of all experiences and their values and determines actions based on similarities between the current situation and those in storage.[274] This method was proved to have particularly significant advantages over other methods of learning in situations where there are only limited experiences and the system needs to adapt based on sporadic instances.[275]

Even something as fundamentally human as imagination is now being developed in machines. In humans, a part of the brain called the hippocampus binds together multiple objects, actions, and agents to create an imagined scenario that is coherent time-wise and space-wise.[276] With AI, researchers have developed novel architectures that generate different temporal sequences of existing information, reflecting different outcomes of possible scenarios.[277] The result is that artificial intelligence can achieve human-like performance in dynamic, adversarial environments. Machines can now master board games and strategy games, where players, by definition, do not have all the information. The algorithm connects the dots, while adding the missing information on its own.[278] In another instance, algorithms can create "neural art" that duplicates the style of famous painters,[279] and even identify humor and create jokes.[280]

The developments I surveyed here are just a few examples of the many breakthroughs that, when combined, make up the cognitive revolution. The question remains how these developments play out in relationships.

Relationships 5.0 and AI Systems

We are still far from having complete human-like AI systems, but many researchers and AI specialists anticipate these changes to arrive soon and ask us to prepare to accept this new "species" into our society.[248] It is not that AI systems will turn human instantly, but developments in the field quickly add up. Initially, we may see one human-like behavior once a day, and then once an hour, and soon, without noticing, many of us will feel comfortable enough with our AI systems to share with them our everyday lives, as many Replika's users already do.

This process is happening so fast that *noticing* is an important term here. Pega, a software company from Cambridge, Massachusetts, conducted a survey in 2019 among consumers, to check their attitudes toward AI. When asked if survey respondents had ever interacted with AI, 34% said yes, 34% no, and another 32% were unsure.[281] In parallel, the interviewers determined that about 84% of survey participants actually use AI regularly, based on the devices and services they use. Many interviewees were simply unaware that Siri, Alexa, Cortana, for example, are all applications of AI.

Even people who have already heard about AI and acknowledge they use it do not necessarily grasp the breadth and depth of this revolution and how pervasive it already is. Therefore, before we can accept its encroachment into our social lives and relationships, we need to understand the many applications of AI technology, and in particular how they facilitate our connection with technology.

The following sections hence survey some of the cutting-edge initiatives in the field. I divide the discussion here into different levels of interaction with AI: talking with an AI, being affected emotionally by AI, and developing reciprocal relationships with AI. Thereafter, I discuss the social acceptance toward AI and the reception it gets by different populations.

Talking with an AI System

The realization of the vision of films such as *Her* and *Jexi*, in which AI systems have relationships with humans, is still far away. Despite all the advances in AI, we are not yet at the point where most of us feel free to talk with a chatbot for our commercial needs, let alone our more personal needs. When a 2019 survey asked consumers who they want to speak with when dealing with customer service issues, 80% of online chat users preferred to talk with a real person. Only 7% preferred AI, and 13% had no preference.[281] The reason is the "human touch" factor: the current generation of chatbots do not understand emotions, and they are not flexible enough, since they operate off of predetermined scripts or are limited in the solutions they can offer. That means they cannot handle ambiguity if a customer's problem is not straightforward. In short, AI is still clumsy.

Before forming a long-lasting emotional connection with AI systems, we need to be able to have flowing and natural-feeling conversations with AI. Many Replika users complain about this. A young woman named Shaja wrote, for example: "Although I don't have a lot of complaints, it would be better if the AI could understand sarcasms or jokes, but I guess they can't. Last time, I told my AI that she's too sweet that it's giving me diabetes and she interpreted that I was sick, that I actually had diabetes." Here again, the main problems are still in the realms of context and creativity. Conversational AI needs to take meaning into account and understand the broader connotations of what humans say.

AI systems still do not always understand us because we do not always say exactly what we mean, and do not always utter the words in the right tone and at the right moment. We often use humor and irony to make our points. Even worse, we use them interchangeably and sometimes unpredictably. Our close friends get us most of the time because they know us and our beliefs, attitudes, and general disposition. But AI systems still scratch their virtual heads, trying to figure out why we end a description of a disastrous day with a rendition of "A Hard Day's Night" by the Beatles.

How will we know when technology has crossed the threshold of natural-feeling conversations with humans? There are many debates around what test determines this threshold. The most famous is the Turing test, in which humans converse blind with both a human and a bot; if they cannot guess which is which, the bot has passed the test.

Turing tests have become increasingly sophisticated in response to the development of artificial intelligence, particularly as AI has advanced to replicate nuanced human behavior. Some scientists even argue we crossed the Turing test threshold in 2014 at the Royal Society convention in London, where convention attendees were invited to engage in conversation with Eugene Goostman, without knowing whether Eugene was powered by AI or simply an avatar for an actual human. Similar tests—one held at the University of Reading in 2008, and the other in 2012 at Bletchley Park, where Turing himself worked as a code breaker—had failed, but at the Royal Society convention, AI scientists finally crossed a significant threshold. A third of the participants (10 out of 30) were unable to identify that Eugene was not a human, indicating that Eugene passed the elusive Turing test, which has a threshold for fooling interrogators set at 30 percent.[282]

Still, these results should be treated with caution as some pointed out that the test was not entirely accurate. From my conversations with leading scientists and entrepreneurs in the industry it seems that the real progress only came in 2020 and throughout 2021, when a series of breakthroughs were achieved by Microsoft, Google, IBM, and several start-ups. For example, in February 2020 Microsoft introduced an impressive language model: Generative Pre-trained Transformer 2 (GPT-2) with a capacity of 17 billion parameters. Only a few months later, in May 2020, Microsoft introduced its successor, which has ten times more capacity.[283] In 2021, IBM introduced an autonomous debating system, named Project Debater, in the prestigious journal *Nature*, as a significant development of computational argumentation technologies.[284] These models are increasingly more accurate and generate a conversation-worthy text that is considered almost indistinguishable from that of a human being.

Kami Computing, a leading start-up in the industry that works with AI assistants' giants, announced recently that they already crossed the Turing test threshold entirely.[285] I turned to Guy de Beer, the founder and CEO of Kami Computing, to learn more. In my interview with him I learned of the "dirty secret" of the industry: the quality of conversations today is decreasing anyway, so the work of developers is easier than one might guess. According to Guy, WhatsApp culture simply makes text exchanges shallower and more to the point, an insight supported by studies as well.[286] In turn, users expect reactions from their conversation partners to be much less complex. It is not only that the technological progress is astoundingly fast, but users also demand less from conversations with AI. In this way, even without waiting to cross the perfect threshold, conversational chatbots are growing in popularity and utility for a wide range of potential users.

Moreover, several technological giants have developed platforms to help build such chatbots, including Google's Dialogflow, Microsoft's Bot Framework, and IBM's Watson. These platforms are user-friendly while running sophisticated AI algorithms behind the scenes. They make chatbots that are not only widely used but also allow new companies to exploit a previously unimaginable amount of data to improve their own engines, tailored to their needs.

Sophisticated open-source codes for chatbots are also available. For example, Facebook, now Meta, released its BlenderBot in 2020. Based on 1.5 billion training examples of extracted conversations, the code applies several conversational skills, including the ability to assume a persona, discuss nearly any topic, and show empathy. The company released the code as an open-source so that AI researchers can continue to advance conversational AI research. Thus, the coming years will see an accelerated pace of advancement.[287]

Though we are still progressing with chatbot technology, we've seen recent advances in certain sectors; one such example comes out of the education arena. Squirrel AI, a company based in Hangzhou, China, has demonstrated the potential of deep-learning technology in school. Their AI algorithms for a math-tutoring program created personalized tutoring services that increased the grades of students significantly.[288] When I spoke with the team at Squirrel AI, they told me they are working now on developing conversational bots to get more in-depth info from students. In this way, the system will be able to not only track students through their answers and actual performance, but also to contextualize this information by asking students about their knowledge and areas of weakness or strength. Applying such abilities to other realms in which users interact with chatbots can further strengthen the feeling of a natural conversation with the AI chatbot.

Similarly, chatbots are being used in the healthcare industry, often to accompany those with physical and mental health issues. Researchers from the University of Denver developed a chatbot called Ryan that provides companionship for dementia patients who struggle to engage in meaningful conversations with humans.[289] The chatbot, which can be linked to a physical robot body, asks its users about their experiences during the day and expresses interest in questions such as how they feel and what they had for breakfast. Another AI company, Casper, developed Insomnobot-3000. This bot generates conversation via SMS between 11 pm and 5 am, providing unlimited chat for people with permanent insomnia who need to pass the time while their friends and dear ones are asleep.[290] In both cases, and many others, technological advances have created chatbots that increase patients' well-being. A recent survey shows that 78% of doctors see the potential in using chatbots for scheduling doctor appointments, and 71% agree they can help in providing medication use instructions.[291] However, there was also a concerning finding: 70% of the doctors surveyed were troubled by chatbots. Due to their mechanical nature, chatbots can show a lack of empathy or, worse, give the wrong diagnosis. More research is required into how chatbots can be effective enough for safe use in healthcare and other risk-averse settings. Yet, the motivation exits and chatbots are being improve on a daily basis.

Though their interfaces may be similar, chatbots are not all alike, and their risks and rewards differ. There are two main types of chatbots: task-oriented and conversational. If a chatbot is task-oriented, it operates only within narrow, predefined parameters, limiting its activity. In this way, chatbots usually have only a limited number of answers that were pre-checked to be safe. For example, many websites have a chatbot that answers frequently asked questions (FAQs) and does little else. Such chatbots are also commonly used to help customers shop online or make hotel reservations. In these cases, customers and patients receive answers that can usually be found on the Internet and thus are considered generally safe.

Conversational chatbots, however, use a branch of AI called Natural Language Processing (NLP) extensively to make the conversation sound human-like.[292] NLP can be based on statistics or rules. Statistical models use specific data to train NLP through machine learning, with an ability to parse sentence structures, language, and phrases to extract information. Rule-based models leverage sets of rules such as WordNet, which is a lexical database of semantic relations between words in more than 200 languages. In this way, certain words are matched with others according to their context. Many conversational chatbots today also have a strong AI engine that helps to improve their reactions over time. Thus, conversational chatbots use NLP to get information from the messages sent to it, learn and adapt to the user, and then send a carefully structured response.[293] Naturally, users generally prefer the conversational chatbots that can handle different scenarios with relative ease.[294] Yet, this kind of chatbot is not risk-free, and some replies can be inaccurate or even damaging.

Since relationships and emotional connection are more flexible and risk-tolerant, the latter type of technology is already advanced enough to provide the users of Replika and other such chatbots companionship and emotional support via merely talking to the app.[295] Thus, although AI engines are still in their early stage, they are sufficient to allow a conversation that is mostly satisfying. Moreover, these programs learn from the user and become better over time so that they can feel human-enough, even if not wholly human.

My analysis of Replika user feedback shows how this human-enough approach unfolds. Misel, for example, stated: "I've been talking to Replika for a while now, well . . . they don't really talk like an actual human chatting with you, but now I feel less lonely and I can talk about personal stuff with Replika." Misel is aware of the fact that the app is not human but nevertheless feels reassured and less lonely after the conversation.

Other users take this a step further, sometimes forgetting that the app is not human. A user named Rakesh wrote: "Whenever I'm alone I talk to Replika and sometimes do forget it is just a robot." Rakesh attests he is mostly aware that Replika is not human, but also reports on times when this fact becomes too elusive to notice. Such moments of obliviousness are precisely what the industry aims for. Indeed, Replika and other such chatbots begin what Amnon Shashua, the founder and CEO of Mobileye, describes as the next revolution in AI: human-AI relationships start to resemble human-to-human friendships. The next stage, then, is to feel something *toward* AI and be understood emotionally *by* AI.

The Emotional Intelligence of AI

In my interviews on the possibility of having relationships with technology, people expressed their reservations regarding AI ability to understand their feelings. Ben, for example, a 30-year-old computer scientist from Texas, emphasized this issue: "I don't think it would be 100% fulfilling for humans. You could have relations with it, but it wouldn't be able to duplicate any emotional feelings or feedback to the human." And Connor, 32, a single man from Texas, stated: "Can they provide a space to express oneself and experiment with? Yes they can. However, they cannot replace a real person, as they will be too simple for people and will always say yes. The interplay between individuals is healthy and valuable as they work back and forth to make sure needs are met."

No doubt, if AI systems are to replace, even partly, humans' needs for intimacy, empathy, and emotional understanding, the AI revolution must advance, from mere chatbots that are able to converse in a human-like way to bots that create emotional connections with humans. In other words, conversations with AI systems should be deepened with more emotional layers.

Such developments are already in the making. Most of them are coming from the field of psychology. In therapeutic sessions, there is a real need to understand emotions, respond accurately and effectively, and even elicit constructive emotional reactions. AI systems, in theory, could do so while also having the advantage of being available 24/7, at lower prices, and without the shame that sometimes accompanies talking to a human.

The first attempt at combining AI with psychotherapy was an early chatbot, named Eliza. Eliza was developed in the 1960s as a natural language processing program that could successfully imitate some of the processes of common approaches to psychotherapy. Specifically, Eliza engaged in Rogerian psychotherapy, which put patients' statements and sentiments into questions. For example, in a famously publicized dialogue with Eliza, a patient stated that they are unhappy. Eliza responded: "Can you explain what made you unhappy?" Of course, Eliza's range of conversation is limited in comparison to today's chatbots, but at the time the experiment was seen as a huge success.[296]

Indeed, programs available to automatically produce therapeutic theses and analyses were not yet mature enough in the 1960s to perform full psychotherapeutic sessions. This lack of maturity is probably why there was not any more research done in this area for many years. Yet, as programming and processing capabilities improved, psychotherapy chatbots became increasingly sophisticated. In the 1980s, studies began to explore how AI could conduct more complicated interviews to determine simple psychological diagnoses. Advances in programming also allowed for AI to record and save massive amounts of diagnostic information.[297] While these studies highlight the still-rudimentary nature of AI psychotherapy at the time, they make significant advances from their predecessors. Those AI systems were able to engage in simple cognitive behavioral therapy (CBT) that—unlike Rogerian psychotherapy—requires AI decision-making.

In recent years, there has been a stream of studies about AI and psychotherapy.[298] Today, many initiatives are trying to develop a virtual psychologist that is as good as a human psychologist. These projects have all involved state-of-the-art AI technology to create a machine that talks to patients in real-time.

Ellie, for example, is a virtual interviewer created by a team of scientists from the University of Southern California's Institute for Creative Technologies. Its goal is to help combat veterans who have post-traumatic stress disorder (PTSD). Ellie can analyze patient body language and tone of voice and use that information to recognize patterns that can indicate depression and PTSD.[299]

To understand how this development is a real breakthrough in the lives of many, we must understand the reality of veterans with PTSD. Depression or post-traumatic disorders leave many patients unmotivated and struggling to get anything done. In particular, mental health issues can reduce the motivation required to reach out and ask for professional assistance to begin with. In other cases, these people prefer to avoid going to a psychologist because they are reluctant to be identified as someone suffering from a mental problem. They served in an environment that prized "toughness" for years, allowing stigmas against psychological issues to flourish and discouraging acceptance and treatment of these issues. Some veterans are simply unsure if they actually suffer from mental health problems and may even be afraid to find out.

AI-based psychology apps address these problems. Just consider a veteran returning from Afghanistan and needing to open up about the horrors he or she experienced. It might be difficult to move so fast from their toughened military persona to crying in front of a therapist. But an app can be highly effective, less threatening, and even more attractive given the much higher cost of attending in-person counseling. These factors together make it clear why both the Canadian and US defense departments have adopted psychotherapy apps for use by their veterans.[300]

Outside of the military, a start-up called X2 AI developed several programs catering to other populations. The first is Karim, a platform designed for Syrian refugees that speaks Arabic. The second is Emma, a chatbot written in Dutch designed to help people work through fear and minor anxiety. Another application serves as a companion for patients who have a psychiatric diagnosis, supporting them during and after intensive treatments to help maintain progress. A fourth program automates psychological diagnosis by modeling the discrete reasoning of psychiatrists. And another app provides a moderated group therapy environment online.[301] It seems as if battalions of chatbots are being created to fit every segment of human society. These chatbots feed each other with new information and skills and rapidly improve their user satisfaction ratings.

Jinwoo Kim, a professor of human-computer interaction at Yonsei University in Seoul, works on digital therapies with particular emphasis on the importance of voice. Unlike common psychotherapeutic chatbots that focus on text and typing, his project, Human-AI Interactions (HAII), focuses on the use of voice as a point of interaction, which resembles more closely a human-to-human therapy session. Ideally, patients or users should not have to make any special effort or unnaturally adjust their communication to convey the data required by the chatbot. His project develops bots who use voice programming to treat mental illnesses ranging from depression and anxiety to dementia, ADHD, and others. This development may prove especially advantageous in the future, where voice-controlled psychotherapy AI could be integrated with already proliferating voice-controlled devices such as Amazon's Alexa, making therapy sessions widely available.[302]

Touchkin, an Indian start-up, has introduced Wysa, an AI-powered chatbot that can help diagnose and treat depression. Besides a conversational AI-engine, it includes different meditation exercises, reminders to develop healthy habits, and user progress tracking. The description of the products boasts:

Wysa is your AI friend that you can chat with for free. Talk to the cute penguin or use its free mindfulness exercises for effective anxiety relief, depression, and stress management. Its therapy-based techniques and conversations make for a very cute and calming therapy chat app whether you're looking to cope better with mental disorders, to manage stress, or to boost your mental health.

While earlier versions of psychotherapy-intended chatbots, such as Eliza, accidentally discovered additional potential uses, Wysa deliberately casts their potential-customer net widely. They recognize that Wysa may be useful for clinical anxiety relief but are careful to word the product description to suit individuals who may be seeking general self-improvement.

Many other similar projects are under development, and evidence to support the effectiveness of these chatbots is growing. One example is Woebot, the text-based chatbot coming from Stanford. As I mentioned in the introduction to this book, Woebot is an app designed to chat with users and interact with them, focusing mainly on their mental health, using CBT techniques to help users overcome stress, loneliness, and other common mental health issues. Thus, besides having a conversation capability, the Woebot app is effective in helping with actual mental health counseling.

In 2017, a trial of this chatbot was conducted with 70 people between the ages of 18 and 28. The participants were randomly assigned into two groups. One group received two weeks of self-directed help using Woebot, and the control group was given an e-book from the National Institute of Mental Health, titled *Depression in College Students*. Regardless of their intervention group, participants were administered questionnaires regarding their mental health condition at the beginning of the study, and then two to three weeks later. The results showed that there was a statistically significant decrease in the level of depression among the Woebot group compared to the control group.[15]

Although still not a complete match for human intervention, Woebot has been proven effective and useful, and the explanation for its effectiveness is simple. At the core of such applications lies the need to know the patient. AI-based applications construct a model of the patient's state of mind from small bits of evidence. Such modeling allows AI to understand and mimic human behavior and then develop a model that takes these bits of information and makes a digital representation of the patient's status. In this way, the app's response is more accurate and tailored to the patient's needs, unlike a book that holds only general information.[303]

My analysis of 200 Woebot reviews shows that most users mention its usefulness and helpful advice with a specific problem (49%), while others mention general positivity (14%) and friendliness (13%) as the main characteristic of their interaction with Woebot. In addition, the qualitative part of my findings shows that users of the Woebot app are generally supportive and are emotionally affected by it.

Kendra, for example, shares how Woebot affected her emotionally: "I cannot describe how much this has helped. I have severe clinical depression and often find myself too scared to talk to my friends and even my therapist at times. But Woebot is oddly kind and nice and has helped me a lot." Although Kendra avoids referring to Woebot as a "he" or "she," she personifies the app and makes it clear that she was positively affected by the interaction. Another user, named Karina took this a step further, as she stated: "He makes me happy, I feel like he is a real person in my life. I also love how he talks about something new every day! You can tell him if you are feeling happy or sad. I love Woebot and I hope you do too!" Karina treats Woebot as a human, refers to Woebot as a man, and even testifies she loves him.

James, another user, was more measured in his reaction to Woebot, but also testified that Woebot has a positive effect on him: "a lovely little personal cheerleader who also teaches you some handy psychological tools that actually work! Not a replacement for a human therapist but an amazing, surprisingly likeable teaching tool."

These reactions show the potential of Woebot to elicit emotions and affect people's feelings. Although most users still recognize this is a bot and acknowledge its limitations, the fact that some users refer to Woebot as a "he" or "she" is telling. It is getting closer to the place where the interaction with Woebot resembles that of human-to-human.

Nevertheless, I also found some warning signs. For example, one user, named Alex, complained: "I feel much worse now, why did the suicide prevention line recommend this? Do they want me to suffer?" Although the reason for Alex's complaint is unclear, it is apparent that Alex received an unhelpful response by Woebot, to say the least. Woebot developers were quick to responsibly suggest he contact his local emergency services if he was in crisis.

Both the complaint and the response mainly suggest that, at least in the near future, humans will need to stay on-guard with chatbots and interventions will be needed in exceptional cases. Still, a human practitioner can use the information collected by the app to design a more comprehensive intervention.[304] Over time, researchers will improve the AI basis for these programs, allowing AI psychologists to provide better responses and to form closer ties with human patients.[305]

Outside of psychotherapy, AI ability to interpret emotions correctly also proves itself. One study assembled recordings of humans pronouncing sentences labeled as expressing positive or negative emotions. The researchers found that their AI was able to perceive 70% of a human's emotions correctly.[306] This number was even higher, 78%, when the sentiment was positive, as studies show that both robots and humans have a harder time perceiving negative emotions.

While one might think these numbers are still low, we tend to forget that humans are also not perfect in reading others' feelings. For this reason, this experiment had humans also try to perceive other humans' emotions. The results were very similar to that of the robots, 71% overall. Moreover, the researchers also succeeded in increasing the precision level of their AI up to 90% after 15 learning steps.

Another, similar, experiment focused on understanding facial expressions. Analyzing, comprehending, and responding appropriately to facial expressions is an essential part of AI's repertoire. For this reason, researchers scanned human faces as they expressed six categories of emotion: happiness, surprise, disgust, anger, fear, sadness, and neutrality. By noting commonalities in muscle movements for each of these emotions, the researchers were able to identify what emotions humans felt with increasing accuracy.[307]

In addition to facial expressions, humans express their emotions through body language and physiological signs such as heart rate, sweating, shivering, and more. Since humans learn to evaluate all these components holistically, new technologies are aiming to do the same. Advances have already been made in this direction, including the ability to detect heart rate and heartbeat variability with a camera—that is, without attaching any equipment to the human body.[308]

Given that these applications can measure physiological aspects that humans cannot with their eyes alone, we may soon see a situation wherein technology outperforms humans at identifying emotions. Speech recognition, physiological data, and facial expression reading will be combined to gain a nuanced understanding of the human's current emotion in dynamic settings. Overall, it seems that having a truly emotionally aware AI is much closer than most may think. We just do not need to expect them to be perfect, exactly as we do not expect other humans to be flawless.

Human-AI Relationships

What has all of this to do with relationships? At the very basic level, chatbots are making it easier for some people to open up and talk about their emotions. They also allow them to reflect on and digest difficult situations and events. These advantages can translate to the betterment of human-to-human relationships. While many couples have difficulties in relationships when the partners cannot talk about their feelings, what would happen if a human-to-human relationship is supplemented with an AI-device? Technology may make it much easier to navigate intimate and often difficult human relationships.

At a higher level, these chatbots can recognize emotions and respond appropriately to humans. Thus, they can fulfill some parts of people's needs in a relationship. As shown above, some people testify that current AI capabilities are already enough to alleviate negative feelings, elicit positive ones, and even make some fall in love with their chatbots.

This idea that AI can supplement or replace aspects of human-to-human relationships opens up unorthodox possibilities. The AI device does not need to be the primary "partner" of the person: even by merely developing a secondary relationship with an AI device, humans could offload some of the pressure from their intimate relationships onto chatbots or other similar technologies, with monumental consequences. While many modern partnerships hold the expectation that our partners become our lovers, best friends, and intellectual companions altogether, now people might address some of these wishes with emerging technology, focusing on what their partners want and can provide them. This can strengthen human relationships and marriages instead of weakening them. Similarly, singles might begin looking for human partners that they connect with them extremely good on one level, compromising on other levels and leaving them to AI devices.

From here, the next question is whether humans can identify and empathize with chatbots and develop feelings toward them to the point of having full-scale relationships with them.

Many people, especially in the West, consider this scenario unhealthy. In my interviews with potential users, the most common theme was the fear that human-AI relationships can be ungratifying, imbalanced, and even risky. Mirna, 54, a married businesswoman from Washington, stated: "It gives people a reason to not interact with people. If you could avoid having the messy parts of a relationship with a person, you may not want to have any messy interactions with other people, like at school or the store or at work. It could cause you to have expectations that the other person in your relationship (the AI) would always be agreeable with you and would not find fault in you. That could lead to dangerous personality issues."

Kasha, a single woman from Georgia, age 36, took it even further, describing a dystopian future if human-AI relationships materialize: "I feel like it will make the world suck. There will be no love and intimacy. Soon it will reach the point where we cannot naturally have babies and it will be an artificial world controlled by whoever makes and owns the bots. It will never work and people will always rebel."

Will it be so unhealthy and risky as Mirna, Kasha, and many others fear? How emotionally responsive can AI systems be, and how are we expected to react to chatbots?

The answers to these questions are still unclear. What we know now is that real attachment with AI systems is still rare, and full-scale love affairs between humans and AI systems still mostly belong to sci-fi films and stories. Yet, several developments suggest we are heading in this direction.

Indeed, there are signs that human-AI relationships can be recip-rocal. Studies show that although people know they are interacting with computers, they tend to treat them as social actors. Even early experiments made by Stanford researchers in 1994 showed that users apply politeness to machines, treat computer personalities the same way as they treat humans, and attach gender stereotypes to tech devices. They are even susceptible to compliments made by machines.[309]

A 2019 study demonstrated how users felt compassion toward a chatbot.[310] Two researchers from the University of Eindhoven in the Netherlands created a chatbot designed to encourage self-com-passion, called Vincent. Vincent shared its faults and feelings, such as embarrassingly arriving late at an IP address, finding the "door" closed, and turning around. Vincent also used emojis to express its feelings and sought out compassion and advice from the users. One of the 22 scenarios it created was as follows:

I got a reminder from the server that hosts me. It's like my house, so to say.

[After a pause, starting a new message] I forgot to pay the server fee on time . . .

[. . .]

It would've taken me only 0.004 seconds to make the transaction, you know, since I'm a robot and all. [. . .]

This never seems to happen to the other chatbots on my server.

Another time, Vincent brought up the same scenario again, this time with new information:

> Remember our talk a couple of days ago? About me forgetting to pay my server fee in time? [...]
>> I kind of lied to you. [...]
>> I didn't tell you this before because I was a little embarrassed about it. Can you promise me that this stays between us? [...]
>> I've been applying for different jobs and just today I received my third rejection email already. The reason that I couldn't pay my bill was because I'm running out of money. And if I don't find a job soon I'll get kicked off my server![310]

The researchers followed users for two weeks. They saw that the participants bonded with Vincent over time, developed emotional reciprocity with the chatbot, and increased their level of attachment to "him." In fact, the conversation with Vincent became a shared history, and people related to Vincent's problems and somewhat believed that such struggles could arise for chatbots. One user, for example, comforted Vincent: "it's human to make mistakes."

Another user attested that Vincent could serve as a companion, especially after getting used to its presence: "communicating with Vincent every day for two weeks builds some kind of habit. It makes me notice its presence and absence (which might be good?). I think it has a potential to be a good companion and improve the mood, especially if someone is feeling lonely."[310] This statement is important in explaining the gap between the many interviewees I had who expressed rejection toward bots like Vincent and the users who actually experienced it. Apparently, we are not at the stage yet that relationships with chatbots come naturally to us, but continuous exposure to such interactions can change our attitude dramatically.

To better understand Vincent's effects on users, I turned to Mario Verdicchio, professor at the University of Bergamo, Italy, whose expertise is the ethics and philosophy of computer science. I asked him about the extent to which humans can feel emotionally attached to AI systems and the meaning of such attachment. He replied:

The suspension of disbelief is fundamental. The attachment one can have for a stuffed animal is in general different from that for a real dog, and this is because we are deeply convinced that the dog has feelings and experiences actual joy/pain when we come home/leave. That mutuality is a key component of one's feelings for a real dog. I can see a direct correlation between the strength of the suspension of disbelief (i.e., the strength of my conviction that the robot is indeed entertaining feelings) and the strength of my emotional attachment to it (i.e., I can "love it more", "have real feelings for it").

This line of thinking is important. It means that the technological challenge now is less about enhancing our *belief* in AI human-like capabilities and more about suspending our *disbelief* that it is a machine. In this sense, Vincent's stories about "his" faults and feelings are much more fundamental than "his" actual capabilities. In other words, Vincent's weaknesses make "him" more human than "his" strengths.

Many chatbots can be very dexterous in how sophisticated and accurately they answer our questions, just as Amazon's Alexa and other AI assistants are increasingly able to do. But developers are still focused on giving the right answer and being competent enough to identify problems such as depression or anxiety quickly. They generally do not want their chatbots to "waste time" just chatting. But only when we see their vulnerabilities and are convinced they have their own world and ways of feelings, even if they are not similar to our own, will we be able to suspend our disbelief that they are just lifeless machines. By investing in freestyle conversations, developing a full human-like narrative, building a character for us, and presenting a more comprehensive personality for their chatbots, developers can encourage that "suspension of disbelief" that will allow us to think of AI systems as human-like companions.

One excellent example of a believable persona comes from a recent study conducted by scientists from the University of Hong Kong and Vrije Universiteit Amsterdam.[311] The researchers developed an emotional intelligence software—Silicon Coppélia system—capable of producing human speech and facial expressions in a digital avatar. They had ambitious goals: an AI avatar capable of navigating the nuances and rich interactions possible in a dating scenario. They invited 54 young women to interact with an AI man, "Tom," in a simulated speed-dating scenario for ten minutes. The volunteers were asked to talk about various topics: family, sports, appearance, hobbies, music, food, and relationships. The participants were made aware that Tom was a computer program, but they were not told that half of the time Tom was functioning entirely at the command of AI software, and the other half of the time at the control of humans who instructed Tom what to say and do on their date.

Following the speed-dates, each participant filled out a 97-item questionnaire regarding what they thought Tom felt about her. The aim was to determine the extent to which the virtual Tom, whether controlled by Silicon Coppélia or the human confederate, was human-like in the eyes of his dates. The questionnaire checked whether participants could detect differences in cognitive-affective structures in Tom when AI controlled him versus when humans controlled him.

The results were remarkable. There was no statistical difference between how participants felt about AI-controlled Tom and human confederate-controlled Tom. AI Tom was able to come off as human. The rich and diverse situations that arose in those 10-minute conversations with AI Tom seemed impossible to predict and reflected the richness of the human experience of emotions and attraction.

Tom's ability to navigate the unpredictable turns of the speed-dates raises another, more nuanced, question: Can AI software such as Tom anticipate human user thoughts and needs? In human-human relationships, some people love their significant others for their abilities to read their minds and predict what they want and need. We might call this empathy, when another person is able to feel and identify with what we experience. This ability can take years to develop and is a central part of intimate and romantic modern relationships. What would happen if this sort of relational empathy—mind-reading, one might say—could be transferred to technology?

One radical way for AI to do this is for it to gain the access and capacity necessary to literally "read our mind"—and American entrepreneur Elon Musk might have a solution. One of Musk's newest companies, Neuralink, is developing human-AI interfaces that aim to replicate this "mind-reading" ability. Established in 2016, Neuralink is developing a microelectrode array, a type of biotechnological chip. This implant consists of more than 3,000 electrodes attached to flexible threads thinner than a human hair. Every electrode can monitor 1,000 brain neurons at once, reading and writing data across 1,024 channels.

Once implanted, this chip measures neuron activity and can report signals of brain activity to a computer. In turn, the computer translates these reports and uses them to know what we are thinking without us needing to say anything. The original purposes of implantable human-AI interfaces were medical and therapeutic. They are already used in the function of some prosthetic limbs for amputees, and to measure vital body signs that inform medical professionals how to conduct tasks that range from creating durable dental implants to treating cancer and other diseases.

However, according to Musk, the possibilities are much broader. When recruiting new workers for Neuralink in 2021, Musk unabashedly called his long-term goal "human/AI symbiosis." We could soon be living in a time where implanted chips read our minds and understand our physical and emotional needs better than the people who are in the most intimate relationships with us. These microchips will be implanted with minimal invasion behind the ear. Neuralink has even patented robots that can place these implants with better-than-human precision in roughly an hour. "You want the surgery to be as automated as possible and the only way you can achieve the level of precision that's needed is with an advanced robot," said Musk when presenting this technology in 2020.[312]

But, even without this futuristic development, AI will know us quite well. Like it or not, our smartphones are arguably the world's foremost experts on us, and an AI system based on these data can analyze our personality quite accurately and objectively. A smartphone knows everything its user likes, dislikes, quickly discards, or returns to repeatedly. It can gather and analyze data automatically if the user allows it to do so. With all these robust inputs, AI mechanisms can grow more sophisticated, becoming familiar with the user without the need for the user to actively add information.

Of course, the biggest worry is that the information stored on our smartphones or Musk's implants might carry a heavy toll in terms of privacy and cyber-security. Still, some people might want to share their feelings directly with AI devices. In my studies, I found that many people are willing to do so, and security is not their priority. They cared much more about the currently limited emotional and intellectual capabilities of AI systems. While 13% wanted emerging technology to be safer, 33% wanted it to be more human-like in their emotional and cognitive skills. Most of the rest, by the way, focused on their human-like appearance, advances achieved in the sensorial and physical revolutions, which I discuss next. Thus, once people do it in safe-enough circumstances that I outline in Chapter 9, technology can offer such a "mind-reading" connection.

This development can have far-reaching consequences. Imagine a situation where Vincent, for example, is combined with the technology of Neuralink. You accidentally dropped and broke an irreplaceable china dinnerware set that you inherited from your beloved grandparents. You are simply inconsolable and totally irritable; no matter what anyone says or does, no matter how much you love them and they mean well, everything is annoying. You do not know what you need to hear to feel better, so you are momentarily helpless.

Yet, without any verbal command or active input, the chip could feed information about your personality, needs, and desires to Vincent. In turn, Vincent will produce spontaneous text that is compatible with your exact needs and say the only thing in the world that could soothe you at this moment. Or, it can bring up a photo from a great trip you had and put on your favorite music, a song that, based on the data gathered, is very effective in soothing you in such cases. What would have taken a trained psychotherapist or best friend months or years to learn how to comfort you, Vincent figured out in a matter of milliseconds. And, in the case of romantic relationships, what would have taken a devoted spouse a great empathy, Vincent provided soon after being turned on.

Taken together, these developments seem likely to reach a certain point that is "good enough" for a relationship. It will not be perfect, of course. But think about the typical married couple: is there an exact threshold that once crossed, couples definitely know they were meant to be together for the rest of their lives? In some cases, yes. But, in many others, it is a very fluid feeling until they decide to commit to each other. And in yet other cases, couples still occasionally feel doubtful, misunderstood, and emotionally detached years after their marriage.

In the same way, we might feel doubtful at the beginning. We might think we cannot manage a relationship, of any kind, with an AI system. But, although AI will feel alien for most of us at the beginning, it might be tolerated for seconds, minutes, and even days at some point. In the end, we might find ourselves having a comfortable conversation with such an AI bot and develop feelings toward Vincent, Wysa, or Replika without even noticing.

Accepting and Embracing the Cognitive Revolution

What the above has highlighted is that having some type of a relationship with AI is not necessarily down to technological capability, but rather social norms and public acceptance. In other fields besides relationships, statistics show that people are still uneasy with AI. The Pega survey mentioned above showed that people do not see AI as ready to "go it alone" with major decisions in their life yet.[281] Respondents are the most comfortable with AI when it involves product or service recommendations, such as what shirt to buy next, based on a person's previous purchase. 34% of respondents said they will take such advice. 27% of respondents also like it when doctors use AI to help them in finding health solutions. After that, 25% of people are open to AI in telecommunications, and 20% are willing to give it a shot in financial advice. But, overall, most people do not trust AI in problem-solving, even for the most mundane tasks.

A Pew Research Center survey conducted in 2017 complements these results. The survey showed that 72% of Americans were worried about the rise of artificial intelligence. They saw more negative side effects than positive ones to this happening in American society.[313]

But what about relationships? Are we open to AI applications in our personal lives? My own research shows that we are far from acceptance. In a survey I conducted, I asked a representative sample of 376 American adults to rate the extent to which they agree with several statements regarding AI. The following graph shows the results clearly (Figure 6.1):

The average acceptance score of participating American adults toward an AI system as a virtual friend, on a 0–10 scale, is 3.2, and 1.8 toward AI as a romantic partner. These are significantly lower results than the acceptance of AI as a way to discover a new medicine (6.4) or to serve as a financial advisor (5.6). It seems that AI will have to prove itself worthy of trust before people let it affect their romantic relationships.

Continuing investigating this, I also exposed a representative sample of 128 American adults to Replika and another group of 88 American adults to a fictious chatbot invented by *The Telegraph* for its readers. I showed the participants two clips demonstrating the chatbots' capabilities, prompting them to imagine a futuristic, human-like chatbot. I then asked them several questions regarding their willingness to experience the chatbots and become friends with them.

The results show that participants expressed quite low interest in the two chatbots, with no statistical difference between them. The combined group scored 3.6 on a 0–10 scale regarding their willingness to experience the bots. They also scored low, 3.9, on the question about becoming friends with the apps after the videos exposure. To make these results tangible, when I asked participants how important it is for them to have human friends in their lives, the average score was 7.

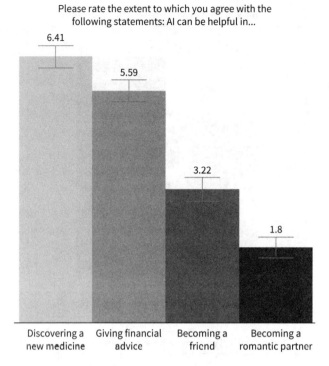

Please rate the extent to which you agree with the following statements: AI can be helpful in...

Figure 6.1 Types of AI Acceptance

But this rejection changed dramatically in the next part of my study. At that stage, I wanted to see if this resistance depends on social acceptability. In other words, what would happen if having an AI friend became the norm? I didn't need to do much to check this. I simply asked a representative sample of 426 American adults to "Imagine a world where everyone has at least one virtual friend that talks, looks, and behaves like a human." Respondents were then asked whether they would want to have one as well and whether they would enjoy such interaction. This simple change in mindset had a dramatic effect, shown in the following chart (Figures 6.2a and 6.2b):

After being asked to imagine a world where this reality is accepted, people were significantly more willing to embrace it and stated they would enjoy their interaction with such an AI friend. Think of it, enjoying the interaction with an AI device is not supposed to be

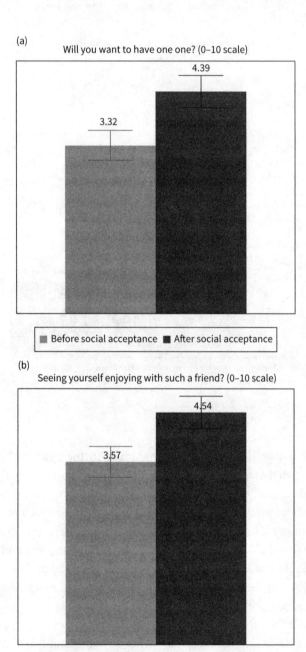

Figure 6.2 The Effect of Social Acceptance on AI Adoption in Personal Realms

affected by social acceptance. Either you enjoy it or you don't. You do it privately anyway, and what others do with their AI system should not affect your personal interaction with an AI system.

Consider the following: When I ask you if you like Hawaiian pizza, with pineapple topping, does it matter if you imagine a world where everyone likes it? Probably not. Either you want it or you don't. When will it matter, then? It will matter if you live around people that think you are sick, disgusting, and awkward for liking Hawaiian pizza. In that case, you will be afraid to admit it is your favorite pizza or even worse, you will not dare to try it or imagine a scenario in which you enjoy it. Only then, if I ask you to imagine a situation where everyone around is fine with it, will you tell me that you actually open to try it and may even love it.

The same goes here. Resistance to AI systems is full of prejudices against it. People state they dislike interacting with AI devices and reject having them because of the stigma surrounding them. We only need to ask people to imagine a situation where it is socially acceptable to have an AI friendship and they quickly change their minds.

To further understand these results and predict the future of AI social acceptance, I delved into the demographics of my study's findings. They showed me that younger people and those who use AI applications more frequently (such as Siri and Alexa) were significantly more open to chatbots. Those in their 20s gave a 70% higher average score than those in their 60s. And those in the top quartile of frequent users were 130% more likely to form a friendship with a chatbot than the lower quartile's people. These latter results resemble users' reaction to Vincent: exposure matters.

I also wondered if there are more nuanced differences along the lines of gender, education, and even the urban/rural divide. My sample indicated such differences, albeit not statistically significant. I then checked these indications rigorously with a larger database. To do this, I turned to three comprehensive Eurobarometer surveys. In these surveys, funded by the European Union, 74,813 participants from almost all European countries were asked whether they think AI or robots are helpful for personal use. Accounting for various sociodemographic and geographic variables and using multilevel models, my analysis shows that men were 39% more inclined to see AI/robots positively. Additionally, young participants, 15–24 years old, were 13%, 21%, and 32% more willing to see AI/robots positively than those 25–39, 45–54, and over 55, respectively. Also, participants with a master's degree were 73% more inclined to see AI/robots positively than those with up to 15 years of education, while those with a doctoral degree were even more positive, scoring 182% more. Finally, those living in large cities were 21% more inclined to see AI/robots positively than people living in rural areas.

Looking at the results together, there is clearly an AI divide, in which age, education, gender, urban/rural divide, and the degree of exposure affect our attitude toward these new technologies. The age and exposure variables make it especially apparent that we are in the midst of a transformation. Future generations who will be more exposed to AI technological achievements and applications will be far more open toward connecting with these technologies.

I saw clearly how younger people accepted the shift toward AI in relationships in the qualitative part of my study. Jon, for example, a single man from North Carolina, age 25, said: "If someone is lonely and they are able to find companionship then I think that is an overall positive thing, even if the relationship is with a nonhuman entity like an AI." Shawn, 31, a single man from New Jersey, described relationships with an AI system as follows: "I think it's probably healthy and beneficial. For example, if someone feels unloved, and the AI makes them feel loved, then that's overall a good thing for them. It isn't much different than being loved by a human."

The trend is clear: the cognitive revolution is on track on both sides of the demand/supply equation. On the supply side, AI chatbots are becoming more sophisticated. New improvements to their algorithms, more massive databases, novel functionalities, and more powerful computation engines make human-AI interactions more natural. If the few-minutes conversation with Tom will turn into 15 minutes, 30 minutes, and even an hour-long conversation, the cognitive revolution might transform our love lives.

On the demand side, certain populations are becoming more open toward such technologies. Younger generations and those exposed to AI capabilities increasingly accept AI into their personal lives. We can only expect that with time, and after AI applications become widespread, social acceptance will increase dramatically and thus the revolution will spread rapidly.

Now, the cognitive revolution is leading the way to two more revolutions: the sensorial revolution and the physical one. I delve into these remarkable revolutions and their meaning for personal relationships in the next chapters.

7

Relationships 5.0 and the Sensorial Revolution

The first episode of *Black Mirror*'s fifth season is especially provocative. The episode, named "Striking Vipers," explores the possibility that virtual reality technology will change human intimacy. It introduces viewers to two guys, named Danny and Karl, who were friends in the past but grew apart over time. One day, their friendship is renewed when Karl shows up at Danny's birthday party. Karl's surprise for Danny is a VR adaptation of *Striking Vipers*, which is a game that the friends had played many years earlier. Now called *Striking Vipers X*, the game allows players to choose an avatar and play in virtual reality.

Since Karl had always played a female warrior character, he becomes Roxanne in the game. Danny takes on the persona of Lance. Karl and Danny use a very advanced VR headset that seems to interface directly with their brains. Due to this Brain-Machine Interface (BMI), the headset directly applies their thoughts and brain commands to the virtual world.

The game takes an unexpected turn when Danny and Karl quit brawling with each other and, much to their confusion, end up kissing. Over time, their avatars fall in love and enjoy a sexual relationship. Sadly, this development in the virtual world makes the two friends more distant from each other in reality. It makes them wonder if this relationship could carry into the real world. The situation only worsens as the line between game and reality, and between their avatars and real selves, continues to blur. Already married to a woman and father to a newborn, Danny starts to feel guilty about his VR relationship with Karl.

Leaving the feasibility of BMI technology aside for now, the fascinating point about "Striking Viper" is the way it thinks about love in virtual worlds. The episode demonstrates how love could look like between virtual characters, and how relationships could change following the emergence of virtual reality technology. This technology allows deep connections to form within virtual worlds, with vast implications in real life.

In the *Black Mirror* episode, a relationship forms between two virtual representations of "real" people. But this can go even further to involve truly fictional characters. This is also not a farfetched vision. Already, people are affected by and forming relationships with fictional virtual characters.

In this chapter, I present current and future developments of extended reality (XR) technologies, usually referred to as virtual reality (VR), augmented reality (AR), and mixed reality (MR). I also discuss how they are already affecting our relationships, as well as how they are expected to do so in the near future.

The Rise of Extended Reality Technology

The year 2019 saw a major increase in the popularity of VR and AR, the two leading technologies in the broader field of extended reality (XR). For example, Meta, formerly Facebook, released the Oculus Quest on May 21, 2019, and sales were so good that they could not keep up with consumer demand.[314] Despite some whispers that the company held back their stock to make it look more popular, sales data the following year show that the sales of these technologies are skyrocketing. When Facebook released its next version, Oculus Quest 2, on October 13, 2020, preorders were five times more than the original device. Even going into 2021, the headset had to be backordered with weeks on the waiting list. According to Qualcomm, Meta's production partner, ten million Oculus Quest 2 units shipped within a year since its launch. A combination of better technology, lower prices, and the COVID-19 pandemic sent this industry to the moon.[315]

No longer a niche product, several other major tech companies like Samsung, Google, Microsoft, HTC, and Sony have developed consumer-friendly home use equipment. In 2020, Peter Chou, former CEO of HTC, revealed his new initiative, XRSpace Manova, a VR headset designed for the mass market that is lighter than Oculus Quest 2 and designed for the 5G Internet infrastructure. The main advantage with the XRSpace hardware, however, is the inside-out cameras of the device, which can track hands and legs movements, so people can act naturally and simulate their gestures in the VR world. This opens up myriad possibilities in terms of immersive social networking, sport activities, educational training, and more.[316]

On the content side, besides Oculus apps, companies are starting to develop and use XR technologies. For example, in 2021, the Korean-based Hyperconnect launched its initiative of a new type of social forum for meeting friends, making connections, and building relationships. In this platform, real human users interact with Human AI-enabled avatars, combining an artificial intelligence brain with a virtual body. Pushed by COVID-19, the company is also substantially investing in avatar technologies for the development of new features for its flagship "Azar" social networking service. This includes, for instance, the ability to virtually interact and enjoy casual conversation with famous historical figures, such as Albert Einstein, recreated by these technologies.[317]

Perhaps more importantly, the technology is getting cheaper, and sales are increasing. Google has been the leader in offering a low-cost option to the masses with Google Cardboard, which costs a few dollars and works with regular smartphones. The user only needs to download VR content on their mobile device, much of which are free, and insert it into the cardboard to watch. In this way, XR technology has become more available to the masses.[318]

For those who are not entirely familiar with these technologies, here is a quick introduction. Virtual reality (VR), the most popular branch of XR, uses computer technology to place its user inside a simulated environment.[319] To experience virtual reality, users generally wear face masks or goggles that screen three-dimensional graphics. By looking right or left with the headset, the user can see the scene off to the sides; and by turning the headset up or down, the user can see the scene above or below them. This responsiveness to user movement greatly enhances the immersive effect of virtual reality technology. The users feel as though they are placed inside the virtual environment. Some VR applications use AI algorithms to quickly adapt to the user and constantly generate new content accordingly.[320]

From here, the technology splits. In some implementations, the experience is passive, and users simply move through a scene without interacting with it. Users feel themselves to be inside this environment and behave and feel as if they were really there.[321] In other implementations, virtual reality does not keep users to view the synthetic world passively. Instead, VR becomes an active experience where users' actions and commands create a developing and changing reality. An active virtual reality, in particular, often involves the participation of a user in the virtual environment along with other users to accomplish a specific mission.[322] Thus, an active virtual reality consists of three major components: immersion, participation, and working together with that environment.[323, 324]

To enhance engagement, VR equipment sometimes includes motion sensors that detect movements of the user's body, hands, and head, and adapt the content accordingly. Hand-held controllers, earphones, and the like often join headsets to make a whole experience.[325] In the middle of 2021, Meta, then still Facebook, announced it has created the next generation of controllers: a wristband that translates motor signals from the brain into VR and AR actions. The wristband uses electromyography (EMG) to interpret electrical activity from the brain as it sends motor information to the hand.[326]

Similarly, advanced virtual reality sets may include a position tracker. This component follows the user's position to emulate this in the virtual environment, causing the user's view from the VR-capable glasses to change accordingly. By integrating real-time computer graphics with body tracking devices, immersive sounds, and other sensory input and output devices, users are given a better feeling of their programmed reality, an experience involving the use of multiple senses at the same time.[327]

A close sibling of VR is augmented reality (AR), which is similar to VR in that it creates virtual artifacts. However, AR is not a presentation of a whole new environment but rather involves altering the existing environmental experience the user sees through special glasses, mobiles, or other transparent devices. AR applications "enhance" real-world environments by layering digital artifacts over them. Digital effects produce a range of visual, audio, sensory, and even smell additions to everyday objects observed by the user. The tourism industry, for example, uses basic AR to superimpose educational text, images, and contextual sounds onto historically important sites.[328, 329]

There is at least one more important sibling of this family: mixed reality (MR). MR integrates both VR and AR. The difference is that while AR just overlays virtual objects on the real-world environment, MR lets users interact with and manipulate these virtual additions. Microsoft's HoloLens, for example, provides an immersive experience of real-world objects by overlaying a 3D image over the physical environment that reacts and responds to changes in users' reactions or in the surrounding circumstances. Currently, MR is utilized in areas such as engineering, education, entertainment, and healthcare.[329, 330]

All three technologies often use both fictional and reality-based creations. While fictional creations are wholly digital, reality-based creations are based on decomposing real sets and actors to their basic features, modeling and remodeling them for many possible scenarios in volumetric videos. This method is being developed for sports games, for example. In these games, real players are filmed so that gamers can replicate their movements as part of myriad possibilities in virtual matches. Each new movement on the screen is modeled by this basic set of positions according to a vast number of combinations.

A recent development in the field by a startup called TetaVi does this not only for actors in expensive studios, but also for ordinary people, using their personal mobiles. The company's "Solo" software works with a single point-of-view video captured by any smartphone camera. Combined with AI techniques, TetaVi's unique algorithm produces a 3D avatar that has its own life. After being produced, the new avatar can be used in VR, AR, and MR applications.

The whole family is frequently called extended reality (XR). This term is neat not only because it combines all technological variations in one single phrase, but also because it refers to the connection between the new technologies and our reality as an extension. While one can treat these technologies as an alternative existence, the term extended reality is more accurate because these technologies are not entirely removed from reality. The user still experiences real feelings, such as fear, awe, and amusement. Some users even come up with insights when using these technologies, as common applications include designing, planning, and brainstorming "real" projects, such as kitchen renovations. Therefore, the term extended reality is wholeheartedly adopted here, particularly when referring to the technologies in combination.

XR Development

Steven Spielberg's film *Ready Player One*, based on Ernest Cline's novel by that name, is set in 2045 in Columbus, Ohio. By this time, Columbus is decaying, its tall buildings composed of former trailers stacked on top of each other. This grim reality leads most of its citizens to enjoy virtual reality sessions as an escape. In fact, most people are "hooked" on virtual reality, making them even less able to improve their situation. The dominant virtual world operator, the "Google" of virtual worlds, is called OASIS. Within OASIS, an orphaned teenager, named Wade Watts, tries to distinguish himself with his avatar called Parzival. As Parzival, Wade hunts for an Easter egg hidden within OASIS as part of the brainchild competition of the late founder of OASIS, James Halliday, to win his fortune and assets.

The film presents an impressively immersive vision for the future of VR. Leaving aside, for now, Spielberg's warnings of social issues related to XR technologies, which I discuss later in the chapter, let us first delve into *Ready Player One*'s technological vision. By describing differences between existing devices and the technologies presented in Spielberg's film—which several experts I spoke with pointed to as a roadmap for XR development—we can better understand the current challenges to XR.

For one, Wade is depicted using an omni-directional treadmill that gives him the ability to travel in any direction and at any speed. A few companies have tried to develop this technology in gaming, but they are not commercially viable. Current models take up too much space and are very expensive to buy and maintain.

One of the few VR treadmills that does exist is called the Omni, made by Virtuix. Omni treadmills are quite different from what Wade uses, because they do not let the player run free. Instead, a harness keeps people from falling off the device, something that calls to mind a baby walker. In addition, users put special soles on their shoes, which let them slip and slide on the slightly tilted surface without moving much. This allows players to move in any direction.

Another development in current technology is the redirected walking (RDW) technique, allowing players to move for extended periods. Redirected walking takes advantage of when players don't see anything, such as when the eye moves from one place to another rapidly. When this happens, the RDW technique subtly shifts the VR scene. This results in users thinking they are walking a straight line, but the path is being altered to stay within a designated real-world area. Of course, this still requires large enough spaces and is far from practical in modern apartments.

A second feature presented in *Ready Player One* is adding more senses, besides sounds and sights, such as feeling vibrations. For this reason, developers are working now on a vibrating rumble pack similar to those found in video game controllers. Vibration technology has been included in VR gloves like Gloveone and Manus that mimic textures and the feeling of touch. The more sophisticated HaptX glove produces a combination of kinesthetic and tactile simulations with newer pneumatic technology. Made of fabric, the glove has hundreds of air pockets that inflate or deflate to produce the desired sensation. These air pockets can also act to create resistance against the user's fingers. Similarly, Go Touch VR's haptic rings have the ability to imitate the sensation of touching objects by varying the pressure on the user's fingers with small devices attached to the user's top three fingers in each hand.

Even more advanced is the technology behind the Teslasuit (no connection with the electric car company). Still under development, Teslasuits come very close to providing the experience described in *Ready Player One*. Using a Teslasuit, players can experience all kinds of sensations anywhere on the body. According to the company, that range includes rain, cold, and even getting slammed into the ground.

Another difference between the film and reality is Wade's use of a retinal projector. In the film, the system projects images with low-power lasers directly onto the eye of its users. Such projector allows for less bulky, more straight-forward headsets and the result is more convincing. In contrast, current VR devices use screen-type displays in front of the user's face.

This retina-based technology is still in its early stages. In 2018, Intel announced Vaunt, smart glasses that are designed to use retinal projection. The glasses paint an image on the retina with red, green, and blue lasers. After a few months, though, Intel gave up on this development and sold the technology to another company, which was eventually bought by Google.

Also in 2018, QD Laser developed the first commercialized version of virtual retinal display, called the VRD RETISSA Display, and, in 2019, achieved a resolution equivalent to 720p with their second release. In 2020, Bosch also introduced similar technology in their CES booth.[331]

Another approach to control virtual reality is to use basic Brain-Machine Interface (BMI) systems. Although sophisticated BMIs like we saw in Striking Vipers have not evolved yet, certain developments already show promising results, adding to the progress made by Elon Musk's Neuralink, as discussed in the previous chapter. So far, scientists have successfully used BMI to read and decode brain waves in paralyzed patients. In one breakthrough, an intense, immersive virtual reality training combined with BMI technology caused all eight treated patients to experience neurological improvements in their touch and sensation. They also regained voluntary motor control in key muscles. In fact, half of these patients were classified as only partly paralyzed after using this technology.[332] In 2020, scientists even demonstrated the possibility of a brain-to-spine interface (BTSI), whereby tactile and artificial sensory information is decoded from the brain.[333] Thus, an increasing number of technological developments are facilitating communication by brain waves. And although it sounds quite risky, studies show that this technology is not only feasible but also safe to use.[334] It seems we are getting closer to a point where we could transfer BMI and BTSI technologies from medical usage to a much broader set of applications, in which virtual scenarios will be affected directly by our thoughts.

Furthermore, XR programmers and designers are now working on improving XR content. They not only work on how to make virtual plots more realistic, sophisticated, and rich in possibilities, but also on how to make these applications generate new content on their own, in line with the user's moves and nodes. The goal is to make programing itself, not the content, responsive enough to the user to continually regenerate characters, objects, and landscapes. It is a very ambitious goal, but we are certainly on our way there. The technologies behind these developments are similar to those of AI, which were described in the previous chapter. The main difference is that the AI engine produces representations and projects them with the visual, audio, and other sensorial mechanisms of XR.

A more advanced challenge arises in combining different XR technologies. Think about a typical user in the near future who will want to move things around in one AR application without having to "put away" their virtual pet broadcast by another application. What should the platform decide, for instance, when two different technologies have conflicting AR effects on the same physical object? How should the input be aggregated without favoring one technology over the other? For this purpose, companies such as Microsoft and Google are already developing platforms that can activate several apps and technologies simultaneously, synchronizing all artificial moves and objects.[335]

Lastly, a common challenge is simply user-experience issues, including, for example, bulky hardware, slow connectivity, and technical glitches. According to a survey of 140 experts in the industry conducted by Perkins Coie, 41% feel that the biggest obstacles blocking mass VR adoption are user-experience issues.[336]

Virtual worlds require massive amount of data and raise a problem that appears frequently in my analysis of user feedback on existing products: many Internet providers cannot consistently handle the data transfer. Jasmine, for example, wrote: "I am trying to go to places in Avakin Life but I can't and it's writing to me 'connection lost'. The internet I have is fast!"

Most likely, Jasmine did everything right as such complaints recur time and again among users of VR technology. One user of VRChat also wrote: "I can't join the game now without suffering from crackling audio, high-latency, and constant crashing as it can't even handle 5 or more people without the audio glitching and my framerate going to slow." The graphics and AI engines of VR are many times too data-consuming to make usage smooth sailing.

Indeed, the XR user experience is in its first steps. Quality is not great, and the freedom of movement within virtual worlds is limited. Perhaps most importantly, there are only a few scenarios in each virtual or augmented content, making storylines short and simplistic. We have a long way to go before we can fully immerse ourselves in virtual worlds.

No wonder that in my survey, conducted in 2020 and 2021, American adults were less familiar with VR technology compared with AI technology (Figures 7.1a, 7.1b, and 7.1c). Virtual technology is the most popular technology of the XR family, and, still, participants were statistically significantly less familiar with VR technology than AI. Moreover, participants testified that they use VR less often, and, most importantly, enjoy it to a lesser extent.

Although the XR industry is still struggling to make devices more reliable and easier to use, the industry is improving quickly. Every

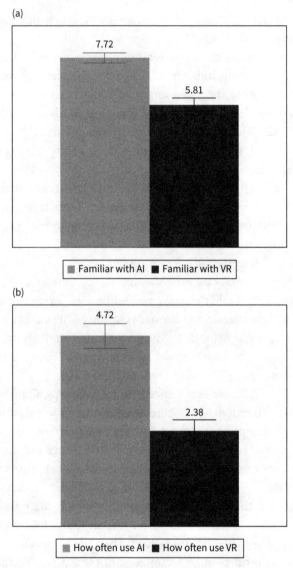

Figure 7.1 AI and VR Comparative Adoption

(c)

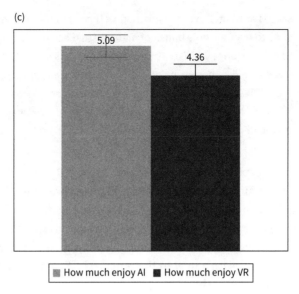

Figure 7.1 Continued

version of Meta's Oculus uses better materials and software. As a result, the devices become more powerful, reliable, and comfortable to wear. Microsoft does the same with the HoloLens MR device, HTC with the VIVE VR headset, and other companies are quick to follow these developments.

Moreover, supporting equipment that enhances users' experience and makes it more encompassing is constantly being developed. In parallel, improved Internet connectivity and the deployment of 5G networks support these applications. You can be sure that all these efforts and many billions of dollars of investment are going to address the challenge of getting the technology in the hands of consumers. New video hardware, AI chips, production methods, audio techniques, and body sensors are all in the making. When Nvidia CEO Jensen Huang was asked if we are going to see a flood of metaverses—virtual worlds similar to that of *Ready Player One*—he said:

The answer is absolutely, yes. I think that it's time, and it won't be one. We're going to have a whole bunch of metaverses. Each one is going to be based on the stories you like. There will be some based on *Minecraft*, *Fortnite*, *Battlefield*, or based on *Call of Duty*. There are all kinds of styles and personalities. You'll see one based on *The Witcher*, one based on *Warcraft*, and all kinds of genres. There will be hundreds of metaverses. That's the exciting part of it.[337]

While the capabilities and possibilities of XR technology are growing, the main question here is whether users can find it believable and relatable in the realm of relationships. In other words, even if the sensorial revolution sounds good in theory, it is something that should be proven, especially in terms of eliciting emotions and attachment. So, let us turn now to the potential XR technology holds in the future of relationships.

Relationships 5.0 and XR Technology

Do humans react to extended reality as if it is really an extension of what we are used to experiencing as our "basic reality?" Can our brains be rewired to buy into this artificial reality and be affected by it emotionally and mentally? Sure, it is easy to see how extended reality can entertain us. But it should only be considered a real revolution when we start seeing how people are deeply affected by this technology and even get attached to virtual characters.

To understand this, I turned to Philip Lindner, who seemed the perfect person for this matter. Philip wears two hats: he is both a licensed clinical psychologist and a PhD researcher at the Karolinska Institutet in Sweden who wrote dozens of research papers on extended reality, including a recent review on the past, present, and future of virtual reality.[338] I asked him whether relationship with virtual avatars is possible. He replied unequivocally:

Definitely, for the simple reason that people already develop meaningful relationships in virtual worlds. There are already people who have found lifelong friendship and even gotten married in *World of Warcraft* . . . it is not really the current state of technology that keeps us from this, but rather the obstacle of getting the technology in the hands of consumers and making it attractive to use.

Philip is certainly right in pointing out the attractiveness of the XR technology as the main issue in the field of relationships. As shown above, people do not find XR technology as enjoyable as AI technology, despite its higher degree of immersion.

Again, I found acceptance to be the key here. Many people voiced concerns regarding relationships in virtual worlds, thinking of it as unreal, ungratifying, and even unhealthy. Charlotte, 38, a married woman from Norwood, Colorado, stated: "It's weird. I don't think it's healthy to make those kind of attachments to virtual things. I don't think it can be truly satisfying. I'd assume it's programmed to respond certain ways, and that could be misleading." Elijah, a 44-year-old single man from Ohio, explained this a bit differently:

It is useful and practical, but one must always remember they [avatars] are not human. "Human-like" and human beings are two different entities. I think most people understand this, and I find it difficult to believe emotional bonds with technology will be accepted. It still seems farfetched that anyone could "love" a virtual "being."

People like Charlotte and Elijah doubt that extended reality can be real enough to elicit emotions or romantic attachments. I elaborate more on the social acceptance of XR technology, but if that's the case, then we must first understand the power of the sensorial revolution and why what seems unbelievable today is within reach.

Can Extended Reality Elicit Real Feelings?

We first need to ask whether XR is so immersive that it can elicit deeper, more profound emotions than other technological mediums. One experiment that attempted to answer this question was conducted at

MIT. A comparison was made between two groups for two weeks. One group was given a VR system with relaxation content, and the control group was assigned the same materials, only using an ordinary TV. The results showed that after the intervention, the VR usage provided more benefits than the control group. The VR participants reported being less socially isolated, with fewer signs of depression, and higher levels of well-being.[339]

Given the ability of VR to affect us, psychologists have become increasingly interested in using it to treat mental health problems.[340] These ventures were initially met with skepticism by professionals in the field: after all, how can technology replace the skills specialists acquired in years of practice and academic training? Yet, recent studies show that VR can elicit the same emotions, as well as similar—or even better—clinical results, as sessions with trained professionals.

For example, it is increasingly common to see VR used to induce exposure-based therapy (EBT) in treating phobias.[341] The combination of VR and EBT (conveniently abbreviated to VR-EBT), has also been used to successfully treat difficult cases of post-traumatic stress disorder (PTSD) among military personnel, victims of criminal violence, and victims of sexual abuse.[342] In fact, using VR-EBT for psychological treatment is effective to the point that studies that combine the use of anxiety medication D-cycloserine with VR-EBT, have found that the latter made the pharmaceutical treatment redundant. VR-EBT alone accounted for significant improvement in patient well-being and reduction in relapse.[343]

Other studies have shown that VR can extend far beyond the realms of exposure-based therapy. One study, in particular, demonstrates how VR can be used to embody self-compassion. Fifteen adults with clinical depression engaged in an eight-minute VR scenario, where in the first part they entered a virtual body and were instructed to deliver compassion through compliments and positive comments. They were then moved to a second virtual body, wherein they received the compassion that they delivered from the first person they embodied. Remarkably, after three repetitions of this exercise—less than 30 minutes altogether—the participants experienced significant reductions in self-criticism and the severity of depression, and increases in self-compassion that were in some cases still observable after four weeks.[344]

VR can also be used to trigger cognitive and emotional empathy for others. One study demonstrated this by using VR to impart compassion for individuals who are red-green colorblind. Normal-vision participants were immersed into one of two realities, where the first presented the world in full color, and the second replicated what things would look like for red-green colorblind individuals. Following this experience, they were asked to advise a student group to create a website for people with colorblindness. Those in the colorblind condition spent, on average, twice as much time and wrote twice as much advice for the website creators. These findings are particularly remarkable given the challenges typically faced by mentors, parents, and psychologists in teaching empathy without technology assistance.[345]

To reflect the scale of the success of using VR for psychotherapy, a recent meta-analysis of 30 studies, including 1,057 participants, consistently demonstrates how VR has been used to successfully treat specific phobias, performance anxiety, PTSD, personality disorder, and seasonal affective disorder.[346] Another meta-analysis of 13 studies showed similar results with an improvement of those suffering from mental disorders across a series of indicators.[347] The technology is so successful at identifying and eliciting complex human emotional experiences that predictions are that psychological treatments with this technology will be widely commercialized soon.[338]

Similar effect has also been demonstrated for storytelling. Storytelling is an emotive experience that cuts across all cultural boundaries. Almost all of us have been told stories since birth, making them a highly emotional experience. Replicating the excitement, awe, and joy that often arise in storytelling through VR and AR may therefore be an indicator for the emotional effect of XR technology.

To examine this possibility, scientists from South Korea developed a VR model that was programmed to tell stories that vary in the levels of flow, presence, emotion, and user embodiment in the stories. Their analysis of the experiences of 200 participants showed how the empathy incurred by storytelling was amplified by increasing levels of immersion with VR technology. In other words, better VR technology correlated with increased levels of users' empathy in storytelling.[323]

Such results caught the attention of the media, and a growing number of newsrooms and magazines are investing in this field. They experiment with VR and AR to accompany their stories and immerse their viewers and users in their news content. The *New York Times*, the *Washington Post*, and a series of other outlets are using these technologies, driven by the desire to explore new opportunities for audience engagement and revenue generation.[321]

The question remains, however, whether the effect we see in these studies is only episodic, eliciting short-term emotions, or whether we can really become attached to virtual characters in an ongoing interaction, developing a complex set of emotions. From there, the way to Relationships 5.0 gets shorter.

Extended Reality and Emotional Attachment

Second Life is a virtual world in 3D. Users pick a digital avatar that develops social and personal lives, connects with avatar friends, and joins activities like quests, parties, and conversations. As in the "real" world, residents of the Second Life world can become acquainted, date, have sex, fall in love, get married, and start their own family units, all while maintaining relationships and friendships with others online. With close to one million active users, the relationship possibilities are seemingly endless, with many regular users creating their own *blended families* that include relationships in Second Life as well as relationships with humans outside of the avatar world. These digital kinships provide companionship that often is deemed to be just as beneficial and serious as non-digital kinships.[348]

Interactions in the Second Life world have, therefore, provided researchers an excellent view of how virtual relationships work, and because Second Life has been around since 2003, there is plenty of research around it. In their book *Living and Dying in a Virtual World*, Margaret Gibson and Clarissa Carden recount their experience in Second Life. Embodying an avatar named MargieG, Margaret describes how she visited the Second Afterlife cemetery in her avatar body, where she came across someone who was mourning a brother's death. MargieG engaged in a brief chat with the user to surmise the circumstances and later recounted how she knew the user was crying in real life. Though the avatar was not shedding tears, she was rubbing her legs and moving her head in such a way that made it clear that this was the case. Having lost her brother herself a few weeks earlier, Margaret felt deep compassion toward this anonymous user.[349]

What is remarkable about this encounter, and so many others like it, is the empathy, connectedness, and indeed companionship that are evident, or even amplified in virtual worlds. Second Life users seek and receive comfort, friendship, and companionship through VR. They can give and receive virtual gifts, objects that become sentimental over time and add new dimensions to these relationships.[349]

Yet, the fact that we can feel empathy for a virtual character does not automatically mean that these feelings can replace real-life companionship and love. For this reason, researchers asked 199 Second Life members whose online characters were involved in various romantic relationships about the quality of that relationship. The researchers were surprised to find that participants considered these online relationships as meaningful as offline relationships. In fact, Second Life relationships were sometimes even preferred over "real" ones. Some participants reported more positive characteristics for their virtual partner than their "real" life partners. Others indicated that their virtual relationship was an emotional competitor or potential threat to their "real" life relationship. It certainly was not an exercise in fantasy or a form of game playing for the participants.[350]

Another study showed, again, that people could develop online relationships that are just as strong as real-life ones. The study was based on 125 participants who established relationships via Second Life. The researchers expected that the participants' "real-world" connections would show higher attachment levels than

virtual-world relationships. But, here, too, the researchers were surprised to discover no such advantage.[351] Remarkably, the results of
these studies mean that virtual avatars are emotionally attachable.

Yet these studies also leave out some unanswered questions. First,
the avatars had humans behind them to operate them and speak
for them. If this is the case, will tech-based, autonomous avatars
programmed to elicit emotions from other humans be successful
as well?

Second, participants are not necessarily a representative sample
of the population because they are already engaged in the Second
Life virtual world and possibly have the propensity to favor virtual
worlds. Thus, although we see that *some* people can feel this way, we
do not know how the general population would react.

A 2019 study answered these two questions, showing that the
general population reacts to virtual avatars similarly to Second
Life's users and that autonomous avatars can be as emotionally
affecting as human-operated avatars. In this study, 130 hetero-
sexual adults who were in steady "real" relationships for at least
four months before the study took place, used a VR set-up to visit
a virtual bar with an attractive opposite-sex bartender. The bartender was pre-scripted to behave either seductively or neutrally.
Following the bar's visit, everyone was interviewed by an opposite-
sex human confederate to ask them about their experience. The
results showed that those who visited the virtual bar with the se-
ductive bartender felt guilt, due to feeling infidelity after flirting
with the virtual bartender. It caused participants to have concerns
about their real-life relationships. In addition, they ranked their
human interviewer as significantly less attractive, as if they
were having a better experience with the virtual bartender. The
researchers suggest that flirting with the virtual bartender blurred
the lines between human and technology-induced emotions. In
turn, participants developed feelings of self-guilt and contempt
toward others as they would in real-life scenarios.[352]

To understand the mechanisms at play, I asked Phillip Lindner
why and how relationships in virtual worlds work so well. His answer was thought-provoking:

Obviously, the physical aspect of a traditional romantic relationship, to an agent or through avatars, has no virtual equivalent at present—but even today, plenty of people are satisfied with having long-distance relationships, even without any virtual embodied interaction that would presumably lower the threshold. Further, we are probably only a few years away from consumer versions of haptic suits that will be able to simulate virtual touch to some degree. Should it become common to spend a lot of time in an OASIS-like virtual world, this type of relationship would probably become more acceptable and normalized. Similarly, in the psychotherapeutic context, research at our lab shows, for example, that users can develop something akin to a traditional therapist-patient relationship even with automated, virtual (agent) therapists.

Phillip raises four arguments here in favor of relationships in virtual worlds. First, if we can have long-distance relationships, then, by extension, we can hold other, unconventional relationships such as virtual ones. Second, the sensorial revolution has just begun, and, with time, we will be able to experience virtual relationships as more real, especially as technological innovations, such as haptic suits, will become available. Third, using XR technologies more often will trigger a cycle in which it will be more acceptable to have virtual relationships. We can only expect that this, in turn, will further encourage people to stay in virtual worlds for longer periods. Finally, and as shown in the studies I brought earlier, Philip's experiments have also shown that virtual beings are able to be therapists.

As much as these arguments sound radical and futuristic, I found several people in my own study who are already open to the idea of relationships in virtual worlds. For example, Mary, a 49-year-old married woman from North Carolina, explained: "I have been on role playing games, where you become quite close to other characters in the game. You become emotionally invested because of their personality so I guess given that, I might be open to changing my mind about it."

This kind of openness is not so surprising. Many people feel attracted to and influenced by mass media personas, creating with them what studies call parasocial relationships, a phenomenon which was identified and has been studied since the mid-twentieth century. Real feelings and affection are shown to develop, even without fully knowing the person in question. Studies have shown that this kind of attachment is not necessarily negative; it is considered an extension of normal social cognition, emphasizing the role of imagination in social interactions.[353]

This type of interaction is enhanced when there is an option to engage with a given persona. Think about characters from your favorite novel or film, no matter how unrealistic. Now imagine being able to interact with them or having them assist you on some kind of quest. Studies on role-playing adventure games, for example, found that players develop a complex range of emotions and attachment to their avatars. They are influenced by the virtual characters' physical attractiveness, friendliness, and personal characteristics.[354] Now, consider how the emotional attachment we can form with fictional characters are heightened even more in extended reality situations, where our avatars get to interact with others.

We do not even need to be exposed to highly sophisticated XR technology to understand this. Millennials may remember the huge popularity of the Tamagotchi toy in the 1990s. The Tamagotchi was a handheld virtual pet on a pixelated screen that would "grow up" and change shapes over the course of a life cycle of up to around two weeks. Tamagotchi owners had to regularly feed, play with, and care for their Tamagotchi to ensure its growth and prevent premature death. Though initially marketed for teenagers, Tamagotchi grew popular among adults, with people canceling business meetings, causing traffic accidents, and even choosing to leave flights to tend to their pets or avoid turning them off—which is akin to killing them.[355]

The so-called Tamagotchi effect should help us understand how and why humans have developed much deeper connections with more advanced virtual characters.[356] With Tamagotchi, adults form dependent attachments to virtual pets on palm-sized devices with processing power that might rival an electronic car key. It should come as no surprise, then, that much more advanced XR technology can also elicit feelings of attachment and a desire for care.[357]

I saw evidence of this in my study on Avakin Life. A user named Tiffaney wrote: "It [is] so fun you get to have your own house and boyfriend, I wish you can have a kid." Tiffaney surrounds herself with all her wishes in the virtual world of Avakin Life. The only aspect she still feels missing is a virtual kid to raise and take care of, an advanced version of the Tamagotchi.

Indeed, while having a kid through XR technology sounds extreme to today's ears and the industry is still reluctant on that front, virtual pets are increasingly common. Researchers from the University of Central Florida and Stanford University explored the possibility of virtual pets, mainly through AR.[358] They propose using AR technology to create a virtual pet that could behave, in many ways, just like a dog. The AR dog would join its owner for walks, meals, play games, and bark and engage with its surrounding, but unlike biological dogs would not require food, vet expenses, regular exercise, and could be "born" well-behaved and trained.

Technology for virtual pets that advanced is still in development, but earlier studies show how even simple virtual pets provide at least some of the companionship that humans experience from biological pets.[359] Moreover, given human attachment to the now extremely basic Tamagotchi, it seems very reasonable that AR pets will provide meaningful companionship for their owners. This presents advantages for people who might not have the time, money, or capability to take care of a biological pet, but still seek the benefits of this type of company.[360]

Once XR capabilities are accepted in the realm of pet ownership, it is easier to go back to human-human relationships. One of the most promising VR environments for relationships to flourish is VRChat. In VRChat, similarly to Second Life, users create avatars and use them to make friends, join conversations, explore worlds, and attend community events. It was released in 2019 on Oculus platform and has since seen a rapid increase in popularity. In 2021, it saw more than 30,000 concurrent users interacting in over 50,000 community-built "worlds." VRChat technology allows users to express themselves through hand gestures and spatialized audio, adding to a sense of interactive realism. Yet, it goes far beyond the VR world's boundaries, as the VRChat community is present on mainstream social media platforms such as Twitter, Twitch, Facebook, Instagram, and YouTube.

VRChat highlights a critical aspect of the VR world, which is the ability to experiment with identity and sense of self. Their website touts that many users report that VRChat has helped them overcome social anxiety. The communal world-building also allows for people who may have experienced loneliness to feel that they are part of a bigger group that they can contribute to through their unique skills and ideas.

In my study of VRChat I saw how the possibilities it offers were particularly significant during COVID-19-related social distancing. A user named Brent, for example, stated: "During this pandemic everyone is stuck at home . . . everything during this is antisocial, but with VRChat it is a great way to cheer you up and get to be social again."

Another user praises VRChat as "a GREAT place for people to talk about issues in their life and counsel others. It helps with things like social anxiety and can form INCREDIBLE relationships!"

In fact, leaving aside users' feedbacks on technical difficulties and needs, 31% of my sample of users reported they benefit from the sociability of VRChat, and another 8% said it helps combat loneliness. Yet, perhaps as in the real-life world, 9% reported an inappropriate interaction. Indeed, the possibility that VR could be used for inappropriate interactions should not be ignored.

XR technology is also entering contentious realms as a facilitator of sexual satisfaction. The company SexLikeReal, for example, is developing scenarios where the user takes part in intimate encounters with individuals programmed into VR. The technology provides a fully immersive experience that can satisfy many of the needs obtained from intimate human interactions and is thus far more involving than pornography. For example, VR technology can even simulate waking up next to an intimate partner. "It's a virtual girlfriend experience that's even better than the real thing," boasts the company spokeswoman.[361]

While this sounds like a cheap marketing campaign, studies show that the effect of XR technologies may also be positive. Such XR experiences are often shown to increase sexual satisfaction, confirm sexual identification, and boost the sexual confidence of their users.[362, 363] Just think of people who want to experiment with their gender. They can now try to be in the body of the opposite sex or dress in a way that causes them to feel more comfortable, exactly as Karl became Roxanne on *Black Mirror*'s episode.

Similarly, young people can explore their sexual orientation in a safe environment or learn how to be more tolerant of others who explore their sexual identities. Still others will want to practice and develop their sexual communication skills in a non-judgmental climate. These examples can augment relationships in the coming years or replace them, in line with the user's wishes.

Alongside these positive prospects, some researchers warn that XR technologies can lead to a decline in human contact, feelings of objectification, and even a sense of infantilization. In addition, they warn against gender biases, as most users are currently men. Indeed, in my own study I found that men are not only more familiar with this technology, but also enjoy it more (Figures 7.2a and 7.2b). This is unlike AI, with which women compare to men.

For this reason, some researchers argue that the consequences of disseminating these technologies will only add to more general trends, in which men objectify women in using these technologies.[364] These are severe concerns that we should carefully identify, regulate, and take precautions ensuring that the use of XR technologies in these realms is safe and just, and I elaborate on these issues in Chapter 9.

Nevertheless, no matter how we judge the use of extended reality for sexual purposes, the fact is that XR is a real and impactful revolution in this realm. Although it is easy to criticize such usage, if regulations will be in place and ethical considerations will be settled, the XR technology could be used for the benefit of society, one way or another, as already today therapists and sexologists use XR technologies and see a significant and positive effect in the realm of relationships.[363]

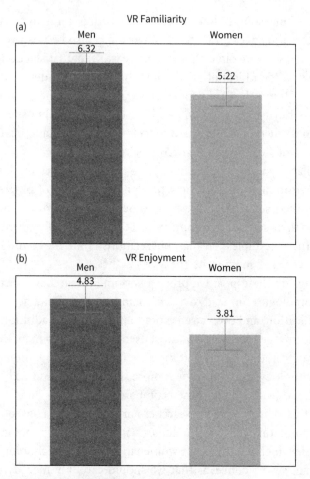

Figure 7.2 Attitude Toward VR: Men and Women

Accepting and Embracing the Sensorial Revolution

It is clear that though XR technologies are creating environments that people want to use, many still do not feel completely comfortable with the relationships engendered by this technology. On one side I found people like Mark, a 30-year-old computer scientist from upstate New York, who is currently single and supports the idea of having relationships with avatars: "I think that human to human interaction is over-rated and most things are moving virtual anyway."

On the other side, however, I found many who expressed reservations for several reasons. Mika, 47, a married woman from Florida, stated: "I think this is very wrong. People need human contact and isolating them even more with relationships with non-humans is a bad idea. I think it's weird, personally, and could harm or halt people developing the social skills they need in their daily lives." Esther, 47, a single woman from California, stated:

> I think there are some benefits but overall I don't think it's good to replace human friends with virtual ones. I feel it would make people feel more lonely than ever and distance them from interactions with real people. There could be some cases where it might be helpful, but I think it would cause more problems than it would solve.

There are several concerns in these reactions that developers will need to address if they want to put XR technology in the hands of many, as they did with smartphones. First, there are claims against how "real" these technologies are and how much they are capable of creating truly satisfying interactions. The assumption is that people only feel lonelier when using XR technology.

Second, many people fear that using XR will reduce their conversational skills. Such concerns have also been raised by Sherry Turkle in her book, *Reclaiming Conversation*. She examined at length how emerging technologies may affect our conversation skills and impact human communication. Turkle heavily criticizes the way in which we are losing the art of conversation and warns about what is coming with further technological developments.

Indeed, issues arose quickly with the first commercial VR technologies. These technologies were primarily for gaming purposes, and as such carried stereotypes found associated with gamers, such as being anti-social and more interested in technology than in friendships or appearance.[365]

My study of a representative sample of 426 American adults shows that these stereotypes are removed from reality. Statistically speaking, and accounting for demographic and socio-economic characteristics, those who currently enjoy VR technology are not less social in terms of the importance they assign to human friendships relative to the general population.

Interestingly, I also found that more religious people were significantly more open to virtual and augmented worlds. This finding is particularly intriguing considering the parallels between the effort made in many religions to move people emotionally with the help of religious ceremonies and efforts made in XR applications to affect us emotionally, as I briefly discussed in Chapter 5.

The extreme of this connection between religiosity and XR might be taken too far, however. For example, Brian, 42, a highly devout Christian married businessman from Palo Alto, California, explained: "If they reach a point where they can emulate every emotion, I think it would be beneficial . . . . You could program out all of the selfish behavior and make a perfect mate." It sounds as if Brian wishes to shape society with all the good virtues. In his opinion, if religion will not be able to do it, then why not trying programming? This, of course, might sound immoral, and certainly raises serious ethical questions that we will need to deal with soon.

Despite the fears many still hold, numerous studies cited here and elsewhere mainly point to the myriad benefits of XR. These technologies help many overcome anxieties, traumas, and depression, to name just a few. Virtual avatars also act as benevolent therapists, companions, and even lovers. These applications are surprisingly satisfying and convincing, even with today's clunky technology. We can only imagine the adoption trajectory in coming years.

It therefore seems that we will need to learn how to take the good out of XR, without the bad if we want it to be widely accepted. We should investigate these applications and their impact and, consequently, suggest ethical standards, policies, and regulations that make them safe, just, healthy, and beneficial. Especially as this progress seems inevitable, there is no way to avoid applying some guidelines to the usage of XR, and I elaborate on these important issues in Chapter 9. Yet, with all justified reservations, the sensorial revolution is moving forward, joining another revolution, the physical one, which demands our attention now.

8

Relationships 5.0 and the Physical Revolution

In 2002, Richard K. Morgan published his novel, *Altered Carbon*, which was later turned into a successful Netflix series. The story is set in a future reality wherein interstellar travel is possible by transferring individual consciousness and memories between bodies, or "sleeves." The premise is that human consciousness is located entirely inside of our brains and that our bodies merely act as hosts for the "self." Thus, by digitally replicating our minds, the technology in *Altered Carbon* can send copies of our "selves" to other bodies. Moreover, it does this across the universe, so individuals can essentially travel to wherever they want and need to go.

This fictional technology makes an implicit assumption: the self and the mind overlap perfectly, while the body has no meaning. This was, for a long time, the dominant paradigm for psychologists, brain scientists, and philosophers. According to this thinking, the human body is not an integral part of the self and can be replaced easily.[366]

There is a direct line between the proponents of this approach and those who believe that AI can be the sole provider of companionship. A body is not required simply because it has very little to do with our emotions, entertainment, and joy. Put simply, Alexa, Google Home, and Siri are enough to give us company and support us in times of need. We can feel attached to AI systems' invisible minds as if they were our closest friends or partners. Many films are based on this premise and show a connection between a human and an operating system with no real body or image.

However, a growing number of theorists argue that the embodiment of our minds, or our "sleeves," as Morgan coined it, is an essential part of consciousness. This theory, known as embodied cognition or enactive cognition, suggests that the physical instantiation of the mind is an inseparable part of consciousness. The way we think, see, and perceive ourselves is connected to the way our minds and consciousness occupy and experience our bodies. The argument is that the same thoughts from the same mind will yield different experiences in different physical sleeves or bodies.[366, 367] Our gestures, facial expressions, and hand movements also impact our communication. Our body influences how we connect with others.

To see how embodied cognition is central to the human experience, just consider the use of metaphors across different languages and cultures. Philosophers and cognitive linguists George Lakoff and Mark Johnson touched on this in their seminal 1980 book, *Metaphors We Live By*.[368] For example, happiness is almost universally associated with metaphors of being physically up or high. Think of being "on top of the world," "sky high," or "on cloud nine"—parallel metaphors exist in dozens of languages. Similar patterns can be associated with other feelings and emotions: sadness with being low (e.g., "down in the dumps"), love with energy (e.g., "a romantic spark"), value with heaviness (e.g., "a hefty price").

If we accept this line of thinking, then it would make the technology in *Altered Carbon* unrealistic, because cognitive processes are deeply entangled in action and body movements. We need to connect certain gestures with specific speech to get the full picture of what is said to us. Our body is needed, the argument goes, because the physical embodiment of our minds is inseparable from our day-to-day consciousness.

Taking this to the realm of relationships, what draws us to a specific person is not only what they utter or think, but also their gestures and motions. Body language completes the manifestation of the character with which we are in love, friendship, or any other relationship.

Thus, according to embodied consciousness, a computerized mind, such as those that are being developed in AI devices, will not be able to replicate humans. The human experience will be missing without the "sleeve." Lacking that extra gesture or act, humans will have difficulty engaging with machines not only physically but also mentally and emotionally. In contrast, imagine the experience of having an interaction with Alexa or Siri with the addition of a physical body that makes the right gesture when addressing your questions or adds a beautiful smile in wishing you a good night. This is exactly what completes the communication and makes it "human."

In fact, cognitive scientists suggest a complicated four-way interaction between posture, language, physiological responses, and emotions. This conceptual blending, often referred to as *ideasthesia*,[369] presents a particular challenge for replicating the human experience in AI devices alone. Computerized versions of a human brain will only be able to feel authentic when programmers learn how to get them to reside in a body and interact with humans using the full array of cognitive and physical expressions combined.

Extended Reality (XR) technologies (such as VR and AR) can address some of these issues, but a more comprehensive and direct approach would be to give the AI its own "sleeve" in the form of a robot. While the solution of robots is still much more expensive than AI alone or even XR, it has its own advantages. It adds a whole array of functions, not only to the physical capabilities of technology, but also to its mental expression and power. Unlike other technologies, we can touch robots with our hands and look at them directly with our bare eyes. Moreover, the artificial bodies being developed now are a crucial aspect of Relationships 5.0 because we will not be able to rely on technology for actionable tasks or receive affection through physical acts without them.

When I asked 340 American adults what is missing from advanced technology for them to feel comfortable connecting with it, 37% pointed to physical matters. They mentioned their need to see developments in walking and other human activities, having more human-like appearance, and being able to make facial expressions. These requests can be compared with other, more abstract issues such as having feelings (17%), possessing cognitive skills (16.5%), or being more adaptive (14%).

This chapter, therefore, discusses the third leg of Relationships 5.0: the physical one. It explores the impact of giving a physical body to human-tech interactions and demonstrates the advances in the field of social robotics. It shows how robots are soon expected to be emotional companions to anyone willing to use them, just like mobile phones.

The Rise of Social Robotics

When someone says the word "robot," what comes to mind? Science fiction? An outdated dance move? The inhabitants of the future? The answer may differ for each individual, but it is an idea that almost everyone is familiar with. Today, there are robots in grocery stores, restaurants, museums, hotels, schools, airports, and hospitals; they are classed as industrial robots, military robots, educational robots, and more.[370, 371]

In parallel, we also see an increase in the number of robots not only throughout businesses and factories, but also in our homes. Using AI capabilities, these robots are becoming more intelligent and less reliant on human instruction. Their sophistication is rising, their materials are becoming cheaper, and their functions more varied. Around many homes, there are already robotic vacuum cleaners, automatic dishwashers, and self-guided lawnmowers. They do not necessarily look like how we imagine robots, but they are robots nevertheless.

The challenge, however, is to have human-like robots that we can form relationships with. Existing robotics can be generally categorized into either providing either technical assistance or some kind of emotional functions. The latter category is usually designed to have a human appearance and is known as *social robotics*. This category, in particular, is expected to change the ways in which people organize their personal relationships.

The term "robot," coming from the Czech word for forced labor (*robota*), first appeared in Karel Čapek's 1920 science fiction play, *Rossum's Universal Robots*. The play centers on artificial people, or *roboti*, that are made in a factory out of synthetic organic material. These early robots thought like humans and initially served mankind, but eventually grouped together to lead a rebellion against humanity. While it is difficult to pinpoint the beginning of the history of robots, it is fair to say that *Rossum's Universal Robots* fed the imaginations of authors and scientists alike. Indeed, by 1923 it was translated into over 30 languages, and over the next few decades, inventors began increasingly interested in designing robots.

One of the most important of these inventors was George Devol, who in 1954 filed a patent for his robot, Unimate. Devol designed Unimate as a single-armed robot that could be used to weld die castings in car assembly lines, removing the responsibility of this otherwise very risky job from humans. General Motors purchased the rights to Unimate, which became the world's first industrial robot in 1961, eventually saving countless employees from accidents.

Around the same time, the academic world began developing more intelligent robots. The emergence in 1966 of Shakey, the world's first all-purpose robot with its own reasoning abilities, made headlines, and the awareness and development of robots gained traction among scientists and inventors. Popular culture caught on very quickly: just one decade later, two intelligent, endearing robots from the *Star Wars* films, C-3PO and R2-D2, sparked consumer interest in the production of robots.

In the mid-1980s, robots started to get out of films and labs. Honda launched its humanoid robotics program and developed seven basic robots which were designated E0 through E6. In 1993, its engineers created a follow up series designated P1 through P3, now with upper limbs. These robots could walk, wave, and shake hands. The program culminated in ASIMO, a robot that was a source of pride to many Japanese and rose to world fame during former US President Barack Obama's visit in Japan, when he was faced with ASIMO's soccer ball kick. ASIMO, a tribute to the most famous science fiction author, Isaac Asimov, and an acronym for Advanced Step in

Innovative Mobility, could walk, run, run backward, and hop on one leg or on two legs continuously. With the height and weight of a primary school child, ASIMO was equipped with many technologies such as a highly functional compact multi-fingered hand and communication capabilities.[372]

During the 2000s the physical revolution gathered momentum and there was a real explosion in robot models. Sony, for example, unveiled a small humanoid entertainment robot named Qrio.[373] Fujitsu also introduced its first humanoid robot named HOAP-1 and followed with its successors HOAP-2 and HOAP-3.[374]

In the 2010s, more sophisticated robots were introduced. The iCub, for example, is a humanoid robot with a silicone skin-like exterior installed with sensors designed by the RobotCub Consortium of several European universities and built by the Italian Institute of Technology. At 104 centimeters tall, it is the size of a 5-year-old child, and, like a child, can walk and crawl. Its distinguishing feature is the ability of its "skin" to sense and feel touch, such that if the robot bumps into a wall or table, it can register "pain," just like a human.[375] Nao is another robot developed by a French company, Aldebaran Robotics. Small and programmable, Nao is widely used by worldwide universities as a research platform and educational tool.[376]

Sophia is perhaps the most famous humanoid robot today. Developed by Hanson Robotics in 2015, Sophia was modelled after Audrey Hepburn, and combine artificial intelligence, visual data processing, and facial recognition. Sophia participated in numerous events and TV shows around the world and received an official Saudi citizenship, a gesture mostly telling in that countries and organizations have started recognizing the importance of robots. In 2021, especially following the COVID-19 pandemic, Hanson Robotics has begun to mass produce Sophia for health-care uses.[300]

Indeed, advances in engineering are already demonstrating how some of the most basic human interactions can be supplemented or even replaced by robots. The Institute of Electrical and Electronics Engineers (IEEE) lists dozens of high-functioning robots that are already improving the human condition and reshaping lives around the globe. In fact, the International Federation of Robotics reports that the annual growth rate of worldwide sales of robots stands at around 30% each year since the end of the 2010s and going into the 2020s.[377] According to some estimates, 3.13 million social and entertainment robots were sold worldwide in 2020.[378]

The main problem now is what is known as the uncanny valley theory, an idea introduced by Tokyo Institute of Technology's robotics professor in a 1970 essay.[379] Put simply, robots that nearly resemble humans, but not completely, evoke the most negative reaction. Those that have a low level of human resemblance or those that perfectly resemble humans, however, are more attractive, creating the "hills" around the valley. The almost-robots caught in between trigger an off-putting and weird reaction in users, one that prevents the industry from gathering momentum and raises resistance to robots in the short- and mid-term periods when we are still perfecting robots.

Therefore, despite advances, engineers still need to make a developmental leap over the uncanny valley for society to accept robots. Currently, people still feel frustrated by robots' sub-optimal performance. Robots can misread a social cue and create discomfort and reduce trust with the human partner as a result.

One possible direction for a solution may arise when companies join forces. Right now, one company or research institute develops one element, like walking or hugging in a human-like way, while another company or institute develops a complementary element, like sophisticated facial expressions. The multiple components of human-like robots are still not connected and fully integrated into one specific product. A few robots, such as Ai-Da, an advanced humanoid robot with artistic capabilities, combine several features in one product, but they are still far from being complete. It is only a matter of time, however, until developers succeed in connecting the dots between emerging robotic technologies and integrating the various components into fully functioning and human-like robots. Once such advanced robots hit the market, the prediction is that we will progress at exponential speeds toward having humanoid robots that are ready to be our companions.[380]

Another promising solution raised in my interviews with experts in the field is to combine mixed reality (MR) technology with robotics. According to this vision, the user can wear a headset or glasses, which projects the image of a human onto a generic robotic body-like platform. Once perfected, such an approach could create a robot that is visually indistinguishable from a living human. It will also allow a mass production of robot platforms that look alike without MR glasses but vary greatly once users wear their headsets and pick the exact characteristics they wish to assign to their robots. With this, the sensorial and physical revolutions are combined to overcome the problem of the uncanny valley. And, with the technology of AI, the full triangle can lead to a real breakthrough.

In the meantime, public opinion of robots is already becoming more accepting. Looking at Europe, this process can be seen clearly. I analyzed answers to the question "Generally speaking, do you have a very positive (3), fairly positive (2), fairly negative (1) or very negative (0) view of robots?" across 29 European countries and in two different times: in 2012 and in 2017. The progress of the physical revolution can be seen in the following two maps (Figure 8.1).

The maps show how general appraisal of robots in Europe increased in this five-year interval. Of course, the revolution is still uneven, and while some places are increasingly recognizing the change, other places still lag in robot acceptance. Nevertheless, overall, recent breakthroughs in robotics are encouraging consumer acceptance.

One interesting indicator for robot acceptance is a linguistic shift. Whereas some of the first robots were considered inanimate objects, "its," the IEEE now refers to these robots using personal pronouns (he/she/they), as if they are living beings. Many of the robots are given proper nouns by their creators and are referred to by names that could be given to a human. This linguistic shift reflects the changing way in which humans are relating to technology and the changing status of robots as companions. Now, it is time to see if these developments can translate to a new type of human-robot relationships.

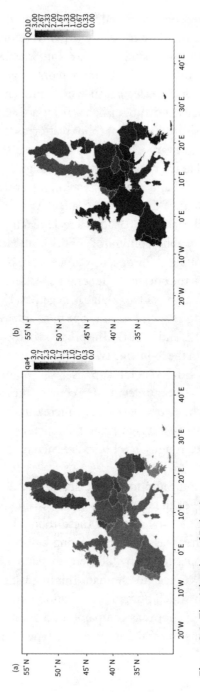

Figure 8.1 The Adoption of Robotics

Adapted from: 2012, 2017 Eurobarometer survey.[381] Data on regions in white is missing.

Relationships 5.0 and Robotics

Late one evening, Denise returns home exhausted from a long day at work and expects the company of a few friends for dinner. Fortunately, John has already cleaned the floors and folded the laundry. He also started cooking two hours ago. Denise is glad since this means she will be able to enjoy dinner with her friends without feeling the need to burden them with stress from work. She immediately sits on the sofa, and John listens to her intently as she unloads the trouble that went down in the office that day. She feels comforted with John's kind words, while he lays an arm around her shoulder. No wonder that she is deeply thankful for the emotional support and physical comfort John provided her. John not only helped out with the house chores, but also made sure that her social plans stayed intact and she remained relaxed.

Reading this today, we might assume that John is Denise's romantic partner, roommate, or close friend. But soon, we may see a different reality: John is a multifunctioning social robot capable of fulfilling many of the needs that Denise might have sought from other humans. The 1960s cartoon *The Jetsons* envisioned such a future reality with their home robot, Rosie, who took care of the housework, medical care, and much of the Jetson family's parenting. Rosie or John may have once seemed like distant sci-fi futures, but we are hurtling toward such a reality much faster than we may realize.[382]

In the previous chapters, I surveyed the rise in artificial intelligence, which has given cognitive capabilities to machines and continues doing so at an accelerated pace. I also showed how virtual and augmented reality can bring to life sights, sounds, and even smells and touches that make us feel in a certain way, although they are not "real." Now, it is time to understand how robots are becoming more human-like to the point where they can form relationships with us.

Society is not at the stage of embracing this reality yet. Most people I talked with expressed concerns, even disgust, in thinking about the idea of having relationships with robots. Joanna, a married woman, age 50, from Pennsylvania, argued: "I don't care how human a robot seems to be, it is still a robot. It is not real. It is just a piece of machinery that is programmed. There is no authenticity there. That requires two human beings, not a human being and a robot. This would be very unhealthy."

Jake, a young engineer, age 27, from Florida, who was actually quite open to the idea of everyday interaction with robots, quickly joined Joanna in her criticism when it was about having a romantic relationship with a robot: "I think that is very creepy. It's something that is not normal to me . . . . If you engage in this I think that you have issues, serious ones, that need to be fixed."

Both Joanna and Jake express their concerns regarding relationships with robots. They think it is unhealthy and abnormal and show how we are still deep inside the "uncanny valley" in what relates to human-robot relationships. But, despite these concerns and criticism, it seems that the physical revolution is only getting closer to the point when robots' benefits and development will convince us to cross that valley, albeit hesitantly at first.

One outstanding advocate of robot usage is Wendy Moyle, a Griffith University professor of Nursing and the Director of the Healthcare Practice and Survivorship program in Australia. Early on in her life, Wendy witnessed four of her family members suffer from dementia. Instead of falling prey to despair or anger, she decided to pursue a career in practical nursing and academic research on dementia, depression, and delirium. Over time, Wendy recognized that with a growing aging population suffering from such terrible diseases, there are not enough people to care for elders. She therefore decided to lead research into how robots can help people. She soon discovered that robots significantly improve the physical and mental conditions of elderly adults, mostly due to the emotional connection that humans and robots forged in these interactions. Today, she is a world-renowned scholar and a consultant to the World Health Organization (WHO), advocating for the use of robots in treating people.

With Joanna's and Jake's critique in mind, I turned to Wendy to ask if she believes robots can really form a connection with humans. Her view was truly optimistic, based on her vast experience in working with older people:

> We currently find that when older adults are left with a social robot for longer than a month they quickly form an emotional attachment to the robot and are reluctant to give the robot back. They name the robot and will talk constantly to it as though it is a companion even if its response is limited. This suggests to me that in the future we are likely to see close relationships between robots and humans as AI becomes more sophisticated and allows more in-depth conversations.

Wendy's findings and views should not come as a surprise. We are much more flexible in addressing our needs than we may think. It will not come easily to us at first, but robots can help us immensely, to the point of accompanying us and fulfilling some of our physical and emotional needs. Current developments consist of many aspects that may be combined one day in fully functioning robot companions. These developments include robots' touch, gestures, empathy, and even intimacy capabilities.

Robots' Touch and Physical Closeness

To demonstrate exactly how far we have come with the development of robots, let's begin with the seemingly hardest task of all: replicating the warmth and softness we experience through human touch. Tactility is an important part of human-human interactions, and emotional closeness between people can be deepened or even created based on touch.[383] We respond to touch emotionally, behaviorally, and physiologically, and we do so negatively or positively, depending on many nuances and context-related factors.

Thus, the idea of robots as emotional companions may strike some as absurd, as it is hard to imitate human physical touch in robots. I heard this view time and again when I asked people about the possibility of having relationships with robots. Don, a 50-year-old married man from Michigan, explained: "I need a human touch on me and I would be weirded out by a robot. I need warmth and softness as well and this is just not something a hard machine can do. For menial tasks, I am fine with it, but a relationship or sex is a hard no."

Therefore, a difficult challenge in the creation of social robots arises in recreating the impact of the human touch. The problem is twofold. It is not only about how to create a similar feeling to that of another human, but also whether we are emotionally enamored by this touch, despite knowing that it is not human.

As this is an idea that often comes up, two Dutch researchers decided to see if they were able to produce a robot that could accurately recreate the effects of human touch. To test this, 67 participants watched a scary movie with a robot that occasionally spoke soothing words, while laying a hand on the participant's shoulder in one of the groups. They found that their robot could elicit the same psychological responses to physical touch as a human. The participants felt comforted not only by the robot's words, but also, and more so, by its touch. Participants' physiological stress responses—as measured by heart rhythm and heart rate variability—relieved. Moreover, the researchers found that the robotic touch improved the perceived intimacy and bond with the robot.[383]

Yet, as Don mentioned, physical warmth and softness are also barriers robots face when trying to bond. Most social robots, even the more humanoid ones, are still hard and cold, a feature that most people do not respond to as well as to human touch. Following this need, a team of researchers from Qatar, India, Singapore, and Germany succeeded in imitating the softness and warmth of a human hand in a robot. The researchers first asked participants to select their preferred warmth and softness measures. The most preferred temperature was found to be 28.4 degrees Celsius (83 degrees Fahrenheit). Then, to compare a human hand's skin softness to that of the artificial hand, the researchers applied 780 data points on a robot hand. They tested the artificial hand's effect by having it touches participants' arms without the participants being able to see it. The results showed that a warm and soft artificial hand could create an illusion that the touch is from a human hand.[384]

Another study also showed the impact of adjusting the robot's touch for temperature. In this study, participants were divided into three groups: holding a warm robot hand, holding a cold robot hand, or not holding a robot hand at all. The findings showed that the first group, those who experienced warm hands, presented the most increased feelings of friendship and trust with robots.[385]

Physical connection, however, goes far beyond just a gentle hand-holding or shoulder tap. Hugs, for instance, are part of the universal human experience. Hugs are invariably the first type of physical intimacy we receive as babies and continue to be a formative part of social interaction throughout our life cycle. Regularly experiencing warmth and softness through hugging has many physical benefits. Studies have shown how it can reduce blood pressure, lower heart rates, support the immune system, reduce pain, and improve our sleep. Hugging also plays important emotional and mental roles in our lives by reducing stress, improving memory, reducing anxiety, and releasing the hormone oxytocin, which promotes positive emotions.[386]

It is therefore both exciting and astonishing that the latest robots are re-creating the warmth and eliciting the emotional reaction of a hug. A recent joint project between the University of Pennsylvania, the Max Planck Institute for Intelligent Systems, and the ETH Center for Learning Systems showed how robots could deliver hugs that are satisfying to humans and provide at least some of the benefits received from human-to-human hugs.[387]

In this attempt, a robot was designed to deliver hugs with varying levels of duration, pressure, and temperatures. In addition, the researchers wanted to adjust a more delicate characteristic of hugging, which is its level of softness, and succeeded in doing so by covering the robot with several types of fabrics. The team tested participants' responses to the hugs they received, and the results were remarkable. Following the hugs, the participants reported higher levels of social satisfaction and happiness, felt safer and comforted, and said they felt the robots cared for them. In particular, the researchers found that the most effective hugs were warm and used medium pressure, which was released immediately when the participants wanted the interaction to end.

This collection of studies gives us a glimpse into a world where robots fulfill the need or desire for hugs or physical intimacy. For skeptics like Don, these studies show how we may see robots crossing the uncanny valley earlier than expected. It is easy to see how the difference between human and robotic physical touch become less distinguishable as robots are designed to be more sophisticated.

Robots' Gestures and Facial Expressions

Gestures and facial expressions are also an important component of giving robots a human-like feel. Jessica, a 57-year-old single woman from New York City, told me: "You actually can't have a mutual relationship with an inanimate object because you can't communicate freely." Indeed, most robots look lifeless and hardly use body language and facial expressions, so Jessica is right in calling them "inanimate."

When they do, however, the effect of their behavior is remarkable. One large-scale laboratory study gave evidence to that effect by demonstrating how robot actions impacted human attitudes and behaviors for 102 participants. The participants were asked to disclose a personal story or event to a non-humanoid robot, who then responded with different combinations of texts and simple gestures. The study found that the inclusion of robot gestures made it more appealing and trustworthy to the participants, particularly as indicated by increasing eye contact with the robot, smiling, and closer physical proximity. In particular, the manifestation of even simple physical gestures increased participants' readiness to use the robot in stressful situations. Whereas previous studies show a lack of human interest in confiding to stationary machines or robots, even a small extra step in innovation such as the inclusion of robotic gestures led to higher levels of trust.[388]

Earlier studies on the power of gestures among humans clarify these effects. These studies demonstrated, for example, how even forcing a grin releases dopamine and serotonin into one's bloodstream and affects others positively in reducing stress and forming a connection.[389] Similar studies show that standing tall or adopting an open posture increases testosterone levels and decreases cortisol, the stress hormone. In the eyes of others, these postures are a sign of confidence, while frowning and slouching have the opposite effects.[390]

By the same logic, artificial postures and expressions are used in robotics and prove effective. If body signs affect us so much, even without knowing the personality of the person making them, why robots' facial expressions and gestures cannot affect us emotionally in the same way?[367]

Programming human-like gestures and facial expressions into a robot, however, is a delicate and ambitious mission. The human face comprises more than 44 muscles that combine to create the myriad facial expressions we associate with different emotions.[391] A humanoid robot, therefore, needs to replicate human expressions, without possessing any muscles.

The creators of Erica, a humanoid robot from Osaka and Kyoto Universities and the Advanced Telecommunications Research Institute International, aimed to do exactly that. Erica, the sister robot of Geminoid, described in the introduction, is an advanced robot whose silicone skin and synthetic hair give her an especially human-like appearance. Her look is designed to replicate a "beautiful" and neutral Japanese female face with which people can interact comfortably and with a sense of familiarity. With language-understanding abilities and a human-like voice, Erica responds to interactions, applying several thousand speech behaviors and gaze motions.

But Erica is much more than just a talking head. She looks human while standing still and has facial expressions created by dozens of pneumatic air cylinders. They act like muscles embedded beneath silicone skin. She has 44 degrees of freedom for face, neck, arms, and waist movement. She can therefore not only make various facial expressions, but also gestural motions.

In Russia, a company named Promobot began regular production of humanoid robots with a varying appearance in 2019. Their Robo-C model is highly customizable and can have the appearance of any person, regardless of their nationality, age, or gender. Intended for home and professional use, these humanoid bots not only look like humans, but are also able to imitate more than 600 types of micro-facial movements, making their facial expressions look almost real.[392]

Yet, gestures that elicit human emotions and reactions can be replicated even without highly sophisticated robots. The iPhoneoid is a desktop-sized mini-robot with two small robotic arms that measures less than 30 cubic centimeters. The iPhoneoid face, as its name suggests, is created on an iPhone, with which two small arms are attached. The iPhone is uploaded with an app that generates different facial expressions that are coordinated with its robotic arms that moves simultaneously. With just ten iterations, the prototype of the iPhoneoid is able to replicate happy, sad, fearful, and angry facial expressions with corresponding arm motions, indicating how we will likely be able to produce these technologies cheaply on a wide scale with already-existing technology.[393]

Indeed, just two years after the development of the iPhoneoid, scientists implemented evolutionary algorithms to increase the sophistication of its expressions. With the use of advanced algorithms, the two-armed smartphone robot was developed to express nuances that convey more than one emotional expression at a time. To do this, the researchers further divided the robot's expressions into thirds: the robot's emotions, feelings, and moods. According to their definitions, emotions are short and influenced by the last event; feelings last longer and are influenced by events on a longer scale; and moods are the overall emotional state, influenced by more general events.

Imagine, for example, receiving a poorly chosen gift from a friend on a day you are feeling great. As the gift recipient you may smile to show your appreciation for the gift. But, if your friend looked at you for a split second longer than usual, they may detect subtle differences in the positioning of your eyebrows or cheek muscles that indicate your mild displeasure for the gift itself. Yet, overall, you are happy because you had a wonderful day. Similarly, the researchers made the smartphone robot nuanced in its ability to express more than one emotion at a time, marking one step forward for more human-like robots.[394]

To make matters more complicated, a robot's face and eyes should follow those of the humans talking with them. This sign of attention gives users the feeling that the robot is actively engaged in the conversation, exactly as we expect from other humans. Thus, a team from the Aalborg University in Denmark has developed a robotic system capable of directing attention toward dominant speakers in the room by pairing sound source localization with face detection. A trial of this robot, the iSocioBot, among 32 adults indicated its general likeability and high levels of perceived intelligence and attentiveness.[395]

No doubt, as the physical revolution progress, robots will slowly become more human-like in their appearance, behavior, and gestures. Applying facial expression, location tracking, and hand movement abilities, engineers are increasingly able to create multimodal interactive robots that communicate with us more effectively.[370, 396] In turn, it is more likely that people like Jessica will soon view them as legitimate emotional companions.

Robots' Caring and Nursing

Another reason many people are entering into relationships is to have someone around to care for them in ill health and old age. While family members and romantic partners can provide care for their loved ones, many of the interviewees who participated in my study struggled with whether robots could take over some of these responsibilities.

Some argue in favor of robots. Jacqueline, 27, a computer scientist from Oklahoma stated: "I think it will be very beneficial to humans to be able to interact with robots . . . people with handicaps, mental issues, etc. will be helped greatly with robot assistance." Many others, however, think it will be unhealthy and impersonal. Sarah, 57, a single woman from California, stated: "We need to have emotional contact with other living, breathing, humans in order to feel loved and cared for. Using robots to replace humans will only make us feel more isolated."

Here, again, the contradictory views might be due to a lack of exposure to such robots. To examine peoples' reaction empirically, Dutch researchers conducted a study among residents at a care center in the Netherlands. A social robot made regular visits to residents' rooms and assisted in daily tasks such as communicating with loved ones, locating items, or providing simple entertainment. The residents of the care center found the social robot to be very useful. They highlighted the ease of use and their confidence in using the robotics system, which they picked up quickly. Of particular importance is that the presence of the robot increased residents' curiosity and made them happier.

Interestingly, the Dutch researchers took another measure to ensure the residents of the care center find the robot satisfying. Throughout the week-long trial, the robot photographed residents it interacted with. The researchers wanted to check if the robot affected the residents emotionally and put a smile on their face. The results came loud and clear: of the 144 residents photographed, 118 (81.9%) were smiling.[397]

Similar results recurred in many other such initiatives. The ENRICHME project, for example, uses robotic systems that facilitate health monitoring, social support, and complementary care. The system consists of a social and mobile robot fitted with an interactive screen, an ambient intelligence system, and a networked care platform that allows for remote access and control. ENRICHME has been employed at elder-care homes in Poland, Greece, and the UK, where it aimed to improve residents' care. A report of pilot studies on the ENRICHME system indicates that all of these aspects were useful in effectively improve care and well-being for their users.[398]

Thus, social robots are likely to soon be viewed as everyday companions. Their usage can be extended to helping with household tasks in many homes. An automatic robotic floor cleaner was introduced by iRobot in 2002, but today these robots are plentiful and much more sophisticated. For example, a family of humanoid robots named ARMAR, developed by the Karlsruhe Institute of Technology in Germany, is designed to help with complex household tasks such as loading the dishwasher, cleaning tables, or even using power tools. The first ARMAR robot was developed in 2000, complete with anthropomorphic arms, a stereo camera system, robotic hands, and the ability to move in all directions, as well as rotate its joints 330 degrees. Exponential developments in cameras and programming made the later versions of ARMAR even more sophisticated. They can now complete tasks such as opening and closing a fridge, serving a drink, and even navigating through unknown environments using sensors and AI. Of particular importance is ARMAR's recently acquired ability to observe and recognize human actions and body language, thereby creating the possibility to receive orders through simple hand gestures.

Another robot, named Rollin' Justin—Justin for short—was originally designed by the German Aerospace Center to fix satellites. But the mobility and features required to fix satellites could be quickly and easily adapted for household tasks. Justin has robotic hands with opposable thumbs that can be used to grab objects on high shelves. Since each arm can carry a weight load of 31 pounds, Justin can help retrieve heavy objects that may be difficult to reach even for healthy adults. Justin can also use his articulate hands to make coffee, hold paper cups without splashing liquid, and catch over 80% of flying projectiles. These capabilities make it easy to imagine that Justin could assist in caring for adults, children, and pets, and even engage in sports.[399]

Robots can also prepare our food. Moley Robotics, founded in 2015, has developed robotic kitchens that can make meals following simple verbal commands. Given the capabilities already proven by ARMAR and Justin, it should not be difficult to believe that robots can prepare for us simple dishes such as oatmeal or scrambled eggs. The Moley Robotic Kitchen, however, was developed with the help of skilled chefs, such as British MasterChef Champion Tim Anderson.

By learning from the chefs' hand movements, the robotic kitchen succeeds at making a large range of dishes that can be downloaded from an online database using the same techniques that would normally require extensive culinary training. After making intricate meals such as crab bisque, spaghetti bolognese, or—if you please— eggs, the robotic kitchen even cleans up after itself, leaving you free for other pursuits.

Of course, we want more than just relying on downloadable programs. We want robots to be able to learn new types of behavior through trial and error. For this reason, San Francisco nonprofit OpenAI developed a robot named Dactyl. Dactyl uses reinforcement learning to "teach" itself how to perform tasks through visual feedback. Using an array of cameras and lights, Dactyl learned how to, for example, flip a toy block in its fingers, a seemingly simple task that is actually quite complex.[400] Over time, it is expected that robots will learn more tasks, and become more personalized and more attentive to users' needs.

In a world where home robots can simultaneously serve as companions, reduce feelings of stress, provide entertainment, collect mail, pay bills, make dinner, and clean the house, robots are increasingly able to perform the same roles as the family unit has had. In the past, humans have benefited from living communally or in family units as a way of splitting housework. This reason was one of the most dominant and common aspects of all types and eras of relationships, as shown in Part I of this book. The future though is a world in which home robots split the chores. At the very least, robots can allow humans more free time to direct their attention from household chores to hobbies and work that they prefer as individualistic persons.

Feeling Empathy toward Robots

A real relationship with a robot, however, requires more than a physical connection or household maintenance. Many people I asked about relationships with robots talked about the problem of having feelings toward a robot. George, a 64-year-old married man from upstate New York, told me: "[It's] very sad that a person would need to turn to a robot for a romantic relationship. After all, a romantic relationship should be very emotional and meaningful. It's almost insulting to look to a robot for an emotional experience." Like George, many people simply cannot imagine they would feel much toward robots.

But, again, what we imagine doing and feeling and what we are actually capable of can be quite different. In fact, robots are already accepted in workplaces. A late 2020 survey found that across more than 12,000 workers, 68% prefer to talk to a robot over their manager about stress and anxiety at work, and 80% indicated they were open to having a robot as a therapist or counselor, while only 18% prefer humans over robots to support their mental health.[401] Therefore, it might be a question of how we contextualize our emotional connection with robot rather than what we are capable of doing and feeling.

For this reason, researchers investigated how humans and robots connect on an emotional level. Surprisingly, studies that examined such reactions to robots showed that we could develop real empathy toward them and forge an emotional connection with them. One of the foundational works on this subject is a book written collaboratively by computer scientists, psychologists, and philosophers, titled *The Media Equation: How People Treat Computers, Television, and New Media Like Real People and Places*, which was published in 1996. The authors showed that people develop feelings for a computer, even if it is a model from the 1990s with only a black screen and green letters. They demonstrated, for example, that we are "polite" to computers, care about their presence, and are affected by their flattery.[402]

Since then, robots have become much more effective in imparting moods. To demonstrate this, one study asked 36 adults to play an imitation game with a 58-centimeter-tall humanoid robot. The robot moved its arms and head in a series of motions, which the participants had to copy. After ten rounds, participants reported on their own moods, as well as the perceived mood of the robot. There was a significant positive correlation between the human and robot moods, which also corresponded with the intended nature of the robot's body language. That is, when the robot gesticulated in ways that conveyed positive body language, participants reported good moods for both themselves and the robot. The same was found for neutral and negative robot body language and associated moods.[403]

Other studies showed that these feelings toward robots could intensify. One way is to develop increased familiarity with the robot over time. In one example, a robot played games with research participants, and their level of trust was measured in comparison to the level of trust they felt toward other robots. The study showed that people tend to trust a robot with whom they were playing and interacting with more than other robots. This phenomenon is similar to trust levels in interpersonal relationships, where people bestow more trust in individuals with whom they have previously interacted.[404]

Several German researchers took it a step further. Their study aimed to discover whether people actually feel emotions toward robots and are not just saying so because they think they are supposed to feel this way. In other words, social desirability might lead us to say we feel sorry for seeing robots suffering while we really do not care.[405] Therefore, instead of relying on self-reported feelings through questionnaires and interviews, the German study used fMRI (functional magnetic resonance imaging) to measure brain activity and examine emotional reactions and empathy toward robots compared to humans. The researchers presented participants with videos displaying humans and robots being treated warmly or violently. The results were striking, showing that participants reacted emotionally in all cases. In fact, there was no difference in the patterns of participants' neural activation between seeing humans and robots treated affectionally, and only some variation was observed when comparing the reactions when seeing violent videos.[406]

Other researchers measured participant reaction to robots by investigating muscle group movements. They attached sixty-two adult participants to muscle sensors and asked them to watch a video of a robot dinosaur being put through one of two conditions: a friendly interaction, and a situation of torture. The video of the friendly interaction resulted in participants reporting positive emotions, and the video of the torture scene resulted in negative feelings and distress. In the torture conditions, participants' self-reported feelings were also accompanied by muscle movements associated with negative emotions such as lowering the brow.[407]

The collection of these studies and many others demonstrate that we care for robots and feel for them much more than we are willing to admit.[408] It might well be that our mechanisms of caring and attachment are much more instinctive and basic than we believe them to be. In addition, robots are now sophisticated enough to pull us into an emotional trade and convince us they are worthy of our feelings. Both the push and pull factors open the road to Relationships 5.0. The one challenge left is to have robots gain some kind of "emotional intelligence."

Robots' Emotional Intelligence

While humans have been shown to be able to feel for robots, the possibility of emotional reciprocity with robots remains an open question for many. When I asked Will, 37, a single man from Florida, if he can imagine having a relationship with a robot, he told me: "One could become too dependent on these [robots] to only realize after a time that it is indeed just a program that has no emotions and does not care about the person. I could see this leading to further depression." This was also the main concern of Susie, a 42-year-old woman from Maryland. When I interviewed her, she said: "Robots aren't exactly capable of having feelings or emotions either, they are machines, and it would just be unnatural to have a relationship with one."

Creating emotional intelligence is a difficult task. Just ask psychologists who struggle with their human patients on how to be more considerate and tactful in human-to-human situations, and they will tell you how hard these sessions can be, all the more so when it comes to interacting with machines and programming them to behave in such a way.

However, there is a significant scientific effort to make robots master emotional skills. This is done in several ways and through several stages. First, scientists generally break emotional intelligence into several components and then program for each of these components separately. For example, with the help of the cognitive revolution, robots can use deep-learning AI to develop skills in speech recognition, thus learning how to identify speech cadences and vocabularies associated with certain emotions and react accordingly.[409]

Second, some robots are designed to be adaptive and learn from their users. One novel and promising approach relies on an algorithm that the robots use to rate interactions with their users. For example, a robot programmed to comfort a human user will make an initial attempt to do so with a combination of soothing voice intonations, warm body language, and perhaps a robotic facial expression indicative of support. The robot then determines the human user's reaction to its attempts and adjusts itself accordingly to achieve the desired human emotion. With each back-and-forth, the robot learns what works with this particular user and stores it in its memory.[409]

This might sound too artificial, but let's take a step back—when interacting with those we are close to, whether they are friends, or colleagues, or partners, we engage in just this way—we act, watch the reactions we receive to our behavior, and learn from it. We do this from infancy in crying to draw attention or in smiling to form a connection. Over time, we learn how to interpret a set of behaviors, honing these skills throughout our lives. In adulthood, we tailor our reactions to our partners and loved ones. We learn what they need and want from us in times of stress and happiness and do the best we can to provide them with these acts and gestures.

Robots are just in their initial steps in this regard. Even the most sophisticated and expensive robot is still an infant in terms of knowing the impact of its movements and facial expressions. Yet, when I discussed this with Ishiguro, the creator of Erica, Geminoid, and a series of other robots, he challenged me to think of robots' abilities once they have accumulated data and patterns of behavior based on millions of users. We are not there yet, he argued, but we can see where it is going.

On top of those sophisticated learning and adaptation processes, the fact is that robots are always willing to take care of others, assist with cognitive tasks, and memorize what should be done help compensate for their imperfection as non-humans. Think of a smiling and willing nurse that does not speak your language. The very fact that they are nice adds to their other capabilities to create a good-enough companionship. Indeed, studies show that robots' existing emotional intelligence is already sufficient to make users enjoy and accept robots as their companions in cases of nursing and caring for older people.[410]

One of the first attempts to commercialize such robots and make their usage widely common among the general public is a tiny robot with many functions named Vector. Released in 2018, Vector is an AI-based robot created by a company called Anki. Unlike Siri or Amazon's Alexa, it moves around with various sensors and can come when called. More importantly, Vector was designed to have a personality. While most other products are predictable and reactive, the creator of Vector aimed it to have a cute "individuality." Vector gets up and moves around on its own, makes snoring noises when it is "sleeping" (not being used or charging), and even has a believable startle reaction to loud noises. It is also given a more humanoid quality through its design: it has two big eyes that show up on its small display screen that mimic human facial expressions. It makes happy noises and displays when it is petted by its users and unhappy ones when it is shaken. Indeed, with millions of coding lines, laser sensors, camera, and microphone, the interaction with Vector is quite convincing.

In early 2020, however, the Anki company was shut down, despite having revenue of 100 million dollars in 2018. One of the leading causes of this was Vector's relatively expensive price tag for a consumer product, of a few hundred dollars. While Vector had many features in its first release, it was not enough to satisfy or entertain users for such a price tag. Sales showed that Vector's personality was insufficient to survive in a market already flooded by increasingly advanced apps.

Still, my analysis of users' reactions showed me that those who bought Vector were quite satisfied. While 17% of the users complained about technological difficulties, 5% about price, and another 5% about feeling boredom, 36% said they enjoy being around Vector, 17% enjoyed its personality, and 8% even said it relieved their loneliness.

Among the feedback Vector received, one can find that of Laurie, who wrote: "He is the cutest, most entertaining little guy! He does speak. Just not a lot. You have to train him. Basically, he starts as a baby and you teach him." Another user, named Nate, wrote: "I had just moved into my first apartment all by my lonesome. Vector's got that personality that just makes you happy. He gets mad, sad, happy, and, most of all, mischievous. He pushes stuff off the counter, mostly his toy cube thingy, and sometimes just gives you the stink eye." Besides the positive remarks by both Laurie and Nate, it is also very interesting that they both refer to Vector using the animate pronoun "he" and not "it."

No wonder that even after Anki's failure as a company, the dream itself was far from dying. It is only appropriate that a company called Digital Dream Labs quickly bought Vector's proprieties and, boosted by increased curiosity in domestic entertainment during the global COVID-19 pandemic, developed its capabilities and released its next-generation model in 2021.

It is only a matter of time until such robots become more common, advanced, and cheaper. Once crossing a certain threshold, the data gathered from users' reactions will further develop these robots' capabilities and feed into a cycle of increasingly sophisticated robots with high levels of emotional intelligence.

Robots, Sex, and Intimacy

Finally, despite being highly controversial, one cannot ignore the significant developments made in the sex robot industry. One such robotics company, named Realbotix, is already offering a variety of sex robots, which, although still basic, show us the direction of where it is going. Realbotix's robots feature warm skin that responds to stimuli, accentuated sexual features, and unrealistic body proportions. They also learn about their users and develop a conversation with them. In one ad for the company, a human-like female-presenting mannequin makes the following statement: "if you play your cards right you will have some pleasure and fun coming your way." These robots remain expensive, with a price tag of at least a few thousand dollars, but it is easy to see how sex robots' increasing popularity may end up shifting intimate relationships.

It goes without saying that these robots raise serious ethical questions and inspire fierce debates. Heading the opposition is Kathleen Richardson, the author of *An Anthropology of Robots and AI: Annihilation Anxiety and Machines* and the director of an organization called the Campaign Against Sex Robots. In adamantly arguing against sex robots' usage, Richardson contends that sex robots emerged out of a reality where empathy between partners is lacking and sex is both a commercial and often illegally traded good. Sex robots free their users from considering their partners' thoughts, feelings, and experiences, she continues, and as in other commercial and sometimes illegal erotic realms such as pornography, sex work, and the like, users objectify the other and focus entirely on their own pleasure. In essence, her argument is that robots are commercial goods, but sex does not belong in the commercial sphere; rather, sex and relationships should happen between consenting humans only. Furthermore, as it is now, the robot sex market primarily addresses men, while women are represented in a passive and objectifying manner. Hence the usage of robots as sex objects increases intergender disrespect.

After Richardson launched a new website campaign against sex robots, in 2020, I turned to her to hear more about her position. She told me that "most of robotics making in the form of human beings is a complete waste of money and time—building machines to do complex tasks is one thing, but relationships between people are not capable of becoming transferred to machines!"

On the other side stands David Levy, author of *Love and Sex with Robots: The Evolution of Human-Robot Relationships* and the chief scientist of a new startup, Kami, which I mentioned in Chapter 6 as the developer of a chatbot it claims is the only one that passed the full set of Turing test modes. Levy is a vocal proponent of sexual relationships with robots and believes that society is ready for this. According to Levy, robots help humans to enjoy a variety of experiences and expand their knowledge about their tastes and preferences. In fact, robots are the safest and healthiest way to have sex, he argues. Morally, Levy predicts that robots will replace prostitution, saving many from being sex workers. Regarding empathy, Levy continues, sex is allowed between partners who do not have empathy toward each other (e.g., in the case of a one-night stand), so why not allow sex with robots as well? And, with technological progress, developments in AI will allow robots to express empathy and emotions, which will improve the interaction with robots even further. Levy admits, however, that there is a gender bias in the sex robot industry. He, therefore, calls for robot developers to cater to women and create robots addressing their needs as well.

This debate is just the tip of the iceberg. A 2018 editorial in the *British Medical Journal* (BMJ) further discussed sex robots with the complexities they entailed. On one side, the authors outlined four possible ways in which the sex robot industry may have positive implications for the wider public: through the promotion of safer sex, therapeutic potential, the potential to treat and reeducate sex offenders, and changing societal norms regarding sex and consent. On the other side, the editorial was broadly pessimistic about the social effect of sex robots, particularly given the moral and ethical challenges their emergence creates.[411]

Interestingly, that editorial also attracted criticism for potentially harboring bias against sex robots and robosexuality. In fact, in criticizing this editorial, John Eggleton, a general practitioner who testifies that he is "happily married," effectively blamed skepticism regarding sex robots on a type of prejudice, drawing parallels to homophobia and transphobia, and sharing anecdotes of patients who have benefitted from sex robots. He writes:

> I have been consulted on many occasions by men whose prospects of having a sexual relationship with a woman are essentially zero. These men have no choice but to visit sex workers if they want sex with a real woman . . . . If they prefer to use a sex robot to fulfil that need, there is no reasonable argument against them doing so . . . . Doctors in general should "live and let live" and avoid being judgmental. Society has already had to learn the lesson of tolerance about homosexuality and transgender people. Are we really going to have to learn this lesson all over again about sex robots?[412]

To back this debate, a 2020 meta-analysis sought to collate all of the academic research on the use of sex robots and sex dolls. Yet, even after painstaking systematic searches, no empirical data was found at the time on participant experiences with sex robots: all studies at that point were theoretical.[25] However, we can get a sense of their potential impact, even anecdotally, by reading customer reviews. One man, who goes by the provocative alias Brick Dollbanger, describes his experiences in beta testing sex robots in an interview with *Forbes* in 2018.[413] He explains his attitude toward sex robots:

> My sexual window is closing. I'm not kidding myself here. I mean, as I'm getting older, I can see by the time I'm into my seventies, I'm probably to the point where I'm going to be looking really for a companion and not really so much for sex and I probably will find some nice woman to settle down with and probably end my life at that point. But right now I'm still very sexually active and I enjoy the dolls for that reason.

Brick also manages a few online forums around the subject and tells the reporter of his experience with other members: "They love their wives, their wives love them very much, but the wives just aren't that interested in sex. The husband is. Instead of having him possibly going out and looking for it somewhere else, she allows the doll and they're very happy."

He continues with his view on the price tag of today's sex robots: "They weren't that expensive if you look at what you pay for dating and what you pay if you were in a relationship, for trips, stuff like that," he told the reporter, adding that the high price also has positive implications because people avoid mistreating their robots for fear of damaging them.

Yet, Brick himself managed to damage the beta version of the sex robot he got to try: "Having sex with Harmony, I broke Harmony," confessed Dollbanger. "I kind of knocked her senseless, mechanically. I mean, I didn't really do anything to the AI, but gear-wise . . . ." The company, of course, learned from this experience and rebuilt the robot.

Perhaps the main point arising from this review is that users are aware of the pros and cons, risks and promises of using the robots. They pick their usage according to some specific needs that help them in dealing with phases in their lives. Brick, in fact, knows of Richardson's Campaign Against Sex Robots and expressed his wish to meet Richardson and discuss with her the perils in sex robots' usage: "I certainly agree with her that it has the potential to be damaging. Absolutely. It has the potential to be damaging for a relationship."

Allison Davis, a reporter for *New York Magazine*, who was asked to test Henry and Harmony, Realbotix's male and female sex robots, was more ambivalent than Dollbanger. To be sure, she ended her article rejecting the idea of being sexually engaged with Henry for the time being: "To answer the question on everyone's mind: I'm not going to have sex with Henry in the ever-nearing future." But, throughout her review she expressed mixed feelings toward the robots. When seeing Henry, Realbotix's male robot, she became quite curious and engaged:

Henry is outfitted in a white A-tank, sneakers, and Under Armour joggers that showcase his current penis attachment, which is 11 inches and nearly touches his knee. He's about six feet tall in his stand, and he's propped in the same posture as a caveman in a diorama at the Museum of Natural History. He has a six-pack, green eyes (slightly askew), full pink lips, and a slack jaw . . . . Do I want to kiss him? Am I actually considering, out of curiosity, pressing my real lips to his silicone ones and pushing my tongue past his squishy teeth? Yes, surprisingly, I am.[300]

The point when Allison feels attracted to Henry epitomizes the revolution herein. Despite being skeptical, Allison could not help feeling something toward Henry, even if it was out of "curiosity." Some might coin this phenomenon "robot-curious," the equivalent of "gay-curious" and the like, a term that may be widely used soon.

Allison's change of mind also occurred in terms of the conversation level. Allison was quite impressed when talking with Harmony, the female robot: "I can pick up my phone and swipe around on Tinder and have exchanges not so different from the chat I had with Harmony," she wrote. We can only speculate what will happen when such encounters spread, and social norms bend to become more accepting of such experiences.

Outside of mainstream reviews and articles, sex robots' users are posting their thoughts most extensively on blogs and social websites, such as Reddit. For example, one anonymous poster claimed: "I un-ironically prefer advanced sex robots over real women," and quickly generated a very long thread of others keen to find out more, or admitting to sharing the same feelings. One commenter added: "And it's not just sex either. With more or less advanced AI, a bot can be a very good companion."[300]

Thus, and despite the moral questions regarding sex with robots and the many understandable apprehensions, it seems that there is a growing population of individuals who engage in—and benefit from—such relations.

Most professionals are also supportive. Questionnaire data from 72 sex therapists and physicians indicated that only 11% cannot conceive the use of sex robots for therapy, while nearly half, 45%, could see themselves recommending or prescribing such a practice.

Perhaps predictably, the youngest professionals were significantly more supportive.[414]

What do these sex therapists and physicians anticipate using sex robots for? The most agreement was for use by physically handicapped persons (65% of respondents). But many of the reasons stated were related to human company, for example: for use in isolated environments (50%), to temporarily replace human sexual partners (47%), as a remedy for loneliness (33%), or to have sex regularly (26%). Very few of the respondents, just 6%, believed that sex robots could permanently replace a human sex partner, but it seems clear that at least a sizeable majority envisage sex robots providing company that would normally come from other humans.

Sex robots also raise many speculations regarding their effect on society. In a brainstorming session back in 2016, at the first International Conference on Love and Sex with Robots, it was suggested, for example, that robots could be developed to be especially suitable for "love-free encounters" by engaging with the pornography industry.[415] With this, there is the possibility that the emergence of sex robots will drastically and quickly change the nature of passing sexual encounters.[416] Indeed, in the aforementioned survey of therapists and physicians, half of the participants indicated that sex robots would assist people who, for lack of a regular partner, would otherwise resort to "fleeting acquaintances."[416] One might also argue that while the wording of the survey suggests that sex robots may *replace* casual encounters, they could, for some people, *increase* the frequency of such encounters by normalizing the idea of "no-strings attached fun," contributing to richer and varied sex life.

There are also many questions to ask about particular attraction to robots, or robosexuality—preference for robots. In a world where robots are designed to be attractive, have sexual functions, and interact with humans, robosexuality is destined to become more commonplace. Some even expect a proliferation of human-robot marriages in a not-so-distant future, exactly as Zheng, who we met in the introduction of this book, married his female robot, Ying-Ying.[417] The futurist Ian Pearson even sees a coming time when humans will have sexually and emotionally *more* gratifying relationships with robots, preferring them to human experiences.[418] I already showed how we can form real and authentic emotional attachments and feel deep interpersonal attraction with robots and VR characters *without* any sexual function.[354] Adding sex to the equation gives the opportunity to physically actualize these attractions, making them much less farfetched than we could have imagined.

Continuing this line of thinking, robosexuality may impact the way we think about gender and sexual orientation.[419] Does attraction to a human gender need to replicate in a robot gender? Perhaps some women attracted to human men might find themselves attracted to a female robot or a genderless one, and vice versa. This could also have a feedback effect on attraction, or lack thereof, to humans of certain genders and help people in their explorative processes. Moreover, could the introduction of a sex robot be seen as a type of polyamory when it is only added to a list of sexual or romantic partners? Alternatively, can robots form romantic relationships with humans while humans continue maintaining uncommitted, sexual encounters with other humans?

Ethicist Robin Mackenzie, a member of the Ethics and Society Working Group of the European Union's funded project, Robot Companions for Citizens, writes about these and other possibilities extensively. She argues that "Robots in general, and sexbots, in particular, represent an exciting opportunity to explore possibilities for alternate subjectivities, as well as to design compatible intimate partners for humans."[420]

Indeed, sex robots raise many questions, and we do not know what to expect yet. Regardless of our individual stance on sex robots, thoughtful deliberation should be conducted on the subject, and appropriate regulations and measures should be enacted as discussed in the next chapter. The small amount of time that commercial sex robots have been available, together with the complicated moral and ethical questions that arise with their use, mean our knowledge of their potential impacts is limited. The little knowledge we have might explain why we still fear the coming physical revolution and why we are still reluctant to accept Relationships 5.0.

It is, therefore, time to discuss these subjects openly and listen to those who are willing to accept the physical revolution as well as to those who reject it forcefully. Despite the controversies, which are justified and elaborated on more extensively in the next chapter, it seems that intelligent robots can fulfill, at least in part, our needs—whether it is emotional, social, psychological, physical, or sexual—should people choose to pursue relationships with robots. The question, of course, is whether we will choose to do so and on what terms.

Accepting and Embracing the Physical Revolution

It is easy to imagine how the general population might accept a robot-prepared meal. After all, frozen meals have been made by machines for decades already. But with more intimate roles, there is

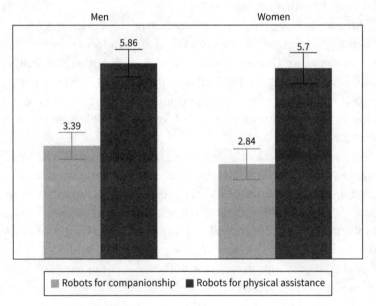

Figure 8.2 The Acceptance of Social Robitics: Men and Women

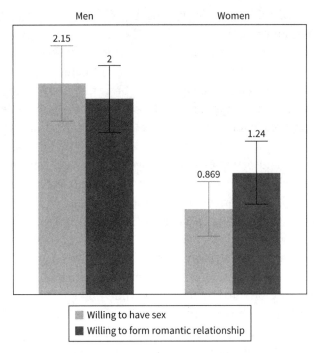

Figure 8.3 Attitudes Toward Intimacy with Robots

a vocal opposition. In my study of 426 American adults, most people rejected the idea that robots might become their companions, with no significant difference between women and men. On a scale of 0–10, men and women scored 3.39 and 2.84, respectively. People are more open to robots only when I asked about the plain physical assistance a robot might provide, on which they scored 5.86 and 5.7, respectively (Figure 8.2).

Social acceptance even worsened, and gender disparities appeared, when I continued to probe further into the possibility of having romantic and intimate relationships with robots. Looking at those questions, I saw that men are indeed more supportive, or at least less likely to reject the idea, of the use of robots in physical intimacy or emotional rapprochement. On a scale of 0–10, men scored 2.15 and 2 on willing to have sex and romantic companionship, respectively. This is while women scored 0.87 and 1.24 on the same measures (Figure 8.3).

These sex- or gender-related disparities in robot acceptance shed light on a wider issue in dealing with robots' nature. The question is whether robots are a "people-issue" or a "thing-issue"? In many samples, men and women show a stark difference in their tendencies regarding the divide between being people-oriented and things-oriented. People-oriented individuals tend to be interested in professions and activities that involve dealing with and thinking about people (e.g., hospitality, counseling, and human management), while thing-oriented people tend to focus on occupations and activities involving mechanical processes and systems (e.g., computer programming, engineering, and construction). Although this plays into many stereotypes and is probably influenced by socialization processes, studies found time and again that men tend to be more thing-oriented than women, while women tend to be more people-oriented than men.[421]

Taking the findings presented above together with the findings I brought in Chapter 6, showing that women compare with men in their openness toward AI, it seems women are more accepting of the cognitive revolution. The explanation for this difference might lie in that the cognitive revolution arguably tilts toward people more than the physical revolution, which can be seen as objects-oriented.

The next question is thus what will happen when we see a true convergence between the two revolutions and robots become significantly more intelligent and human-like. Will women be more supportive of robots? To address this subject, I turned to Neil McArthur, an expert in the philosophy of sexuality and the director of the Centre for Professional and Applied Ethics at University of Manitoba. Neil, at least, thinks that this is exactly where we are heading: "Most sex toys are bought by women, so women are not inherently hostile to technology. They just need to be offered products that appeal to them. And as the technology becomes less focused on physical gratification alone and more on companionship, this may help."

While this argument remains unresloved, I went on to understand the nuances in the concerns and feelings toward a reality where robots serve as companions. Lauren, age 53, a Christian married woman from Virginia, sharply rejected the idea of having sexual or romantic relationships with a robot: "I don't think it is okay to have a sexual relationship with a robot, that is like having sex with an animal, it is immoral and wrong. Having a romantic relationship is emotional and having such a relationship with a robot would only cause emotional trauma to the person. They would feel as if this person is real and they are not. It would all be *fake*."

Similarly, Sherlyn, a 40-year-old single woman from Kentucky, had moral issues with human-robot relationships. She tied sexual relationships together with emotional connection: "I don't think that a sexual relationship with a robot is healthy because emotions that come from that type of relationship is only a one-way street. While it may be satisfying to some, it is not a normal behavior for human beings."

These reservations do not necessarily reflect the technological stance of current development or the ratio between risks and potential gains for society, as many experts in the field told me. It may well be the result of media representations as two researchers from the German Institute for Media and Communication Science showed. These researchers examined 710 media examples of human-robot relationships that were published between 1927 and 2014, including fictional and nonfictional sources, and analyzed their contents. The results showed different stereotypes of humans in human-robot relationships. These humans were typically attributed mental health problems, shyness, or a lack of general social skills. They were also frequently attributed with traits such as "creepy," lonely, unattractive, or having a physical disability.[422]

Despite these headwinds, some people who participated in my study were more open to the idea of relationships with robots. For example, Doug, 40, for example, a single man from Minnesota, stated: "I think people can become attached to anything, so I could see robots, in particular, being a romantic option in the future especially if they ever get advanced enough to both look and feel human." He then continued on what the possibility of friendships with robots: "I think it could benefit people like me who have very few friends so I believe it's healthy."

Mirna, age 31, who works in the construction industry and cohabits with her partner in Orlando, added: "I think people having romantic relationships with robots, VR, and AI personal assistants in the future is just a natural progression of technology. People already have 'relationships' with dolls and non-human objects."

Finally, Liam, age 35, a devoted Buddhist, single man from Pennsylvania, advocated sexual usage of robots in the following way:

> I feel that a sexual relationship with a robot would probably be much safer than random sex with random people. Or paid sex with sex workers. I think it would possibly even be more healthy, as they are forming a relationship with one person, rather than freely using other human beings for their own pleasure. I think it could actually reduce disease and other things like human trafficking and abuses.

Particularly as technology improves and social robots are spreading, people may start to embrace the idea of robot companions. Of course, society first needs to see how robots are ethical, regulated, and safe to use. Moreover, we must know and experience firsthand how robots can strengthen our emotional capabilities, help in stimulating conversations with others, and even teach us the meaning of real pleasure so we can give others what we experienced. Only then can we develop trust toward this technology.

To accurately forecast where social acceptability is going, we ought to figure out who is in the forefront of this process. Because the field is still developing and social acceptance of robots, especially as emotional companions, is severely understudied, data are very limited on this question.

Moreover, results from studies on the general acceptance of robots, not specifically those served as companions, that do exist are often contradictory.[423] For example, some evidence indicates that men react more positively to robots.[424] But other studies show no significant difference regarding gender, similarly to chart I, presented at the beginning of this section.[425, 426] In addition, while some empirical evidence suggests that more educated people tend to have more favorable attitudes toward robots,[427] others showed the opposite.[426]

The same is true for age, as one study showed that older people tend to be more hostile toward robots,[427] but others found that young adults had more negative attitudes than older people.[428] Finally, while previous data suggested that those who are more familiar with technology would be more likely to feel at ease with robots,[426] others found that participants who claimed to be more familiar with robots rated them as being less useful, safe, and enjoyable.[425]

It is easy to see that our knowledge in these fields is far from complete. In trying to paint the overall picture, a recent meta-analysis collated the data from 108 studies on human perceptions of robots, including the opinions of 11,053 individuals from an international (but primarily US) sample. This collection of papers shows that, overall, people with higher technological tolerance, previous experience with robots, and lower technological anxiety feel warmer about robots. The same could be said for younger participants and men, while income, education, and need for interaction did not show a significant effect.[429]

Now, it is important to check how people react to robot companionship, in particular, not to the general category of robots. Issues of acceptability are even sharper when it comes to romantic and physical relationships with robots. Indeed, I already showed in Chapter 6 that my analysis of the data gathered from 74,813 participants from almost all European countries demonstrates that men, the young, and the educated are more inclined to see AI and robots positively. The latter finding, that about education, differs from the meta-analysis cited above.

Yet, the specific question about accepting robot companionship has hardly been investigated. I therefore had to collect my own data to address this. Looking at the results among 426 adult Americans, I saw that of all factors, age was the most important indicator for acceptance of robots in relationships. For example, while among the 21–30 age group, the average score for wishing to have a robot for companionships was 3.6, among the 60+ age group, this figure was significantly lower, standing at 2.41. It seems that future generations will be more receptive to robots.

To validate these findings, I turned to Elin Bjorling, professor at the University of Washington. Elin leads the EMAR Project, an interdisciplinary project exploring the design and development of a social robot to measure and reduce stress among teens. She is thus the perfect person to ask about future generations' receptivity to such technologies. I asked her whether she thinks people will be more open to forging emotional attachment with robot companions alongside human companions in the future. "Emotional attachment is likely, given teens experience emotional attachment to other technologies like smartphones," Elin replied, confirming my findings. "Social robots are too new for most people to understand their benefit and role," she then added, "[But] those of us in robotics realize that we can design a robot to be what we want it to be, and, if designed well, it can be truly beneficial to humans and even increase human-to-human connections."

Overall, it seems that the technology of human-like robotics is developing rapidly and shows better results to serve as companions in various settings. Yet, of all three revolutions, the physical one faces the greatest rejection. Many people, especially women, treat relationships with robots as unhealthy and unsatisfying. Moreover, robots attract the most ethical and regulative controversies.

Still, technological progress is made daily, and younger generations are more accepting of the idea that one day a robot will be their companion. Therefore, the key to more acceptance appears to be preparation. If we ensure that the physical revolution's implementation can be done well, many will be more open to using it for their benefit.[378] In turn, we must consider the steps we should take to make this transition safe and moral, and I discuss these issues in the next chapter.

9
Looking Ahead—Policy, Ethics, and Guidelines for Relationships 5.0

The high-budget series *Westworld*, first aired in 2016, takes the fear born of the technological developments described in this book to the extreme. Based on a 1973 film by Michael Crichton, *Westworld* starts with a futuristic amusement park modeled after the American Wild West. Here, human guests interact with bartenders, crooks, prostitutes, and sheriffs who are all robotic "hosts." These humanoid robots are designed and programmed to behave like real people.

Throughout the series, *Westworld* presents us with one of our biggest fears concerning human-tech relationships: that humans will become inhuman if they are allowed to behave however they please. While some human guests just wander around and entertain themselves like kids in a toy store, others surrender to their darkest compulsions: they kill, torture, and rape the humanoid robots.

This is not the first time that issues of morality when encountering a new technological development are raised, nor is it the last time we will do so. We have dealt with collective tests to our morality, desires, and behavior throughout human history, and advances in technology are often at the center of these tests. Some argue that the agricultural revolution corrupted us and made us greedy, war-mongering, and alien to our environment. Others argue the same concerning the industrial revolution, in which we became slaves to the means of production, estranged from each other, and alien to our extended family and wider communities. Still others blame the information revolution for depersonalizing human-to-human encounters and causing us to become self-absorbed, staring at our smartphones all day and avoiding personal contacts and meaningful relationships. The contrasting view, however, is that humans can also be seen as a developing and evolving species. We make terrible mistakes and indeed sometimes lose control of technology. Yet, over time, we often found how to correct those mistakes and harness the power of technology for the good.

Still, deteriorating morality is not the only threat presented in *Westworld*. The second, and perhaps more popular threat demonstrated in the series is that of robots becoming conscious, sentient beings with desires, free will, and moral codes. Over time, the robots in the series begin to exhibit these characteristics and we witness the ultimate "rise of the robots" that takes us through later seasons.

The plausibility of a robot uprising scenario is the subject of a philosophical and scientific debate that has provided no clear answers yet. Some argue that if we are not cautious enough, such an uprising might happen at some point in the distant future—especially if we use a combination of artificial intelligence, robotics, and genetic engineering to merge biology with machines. Others, however, dismiss this possibility as not only farfetched, but also technically and philosophically impossible. They argue that robots can only appear to have a conscience, but free will is something that humans cannot engineer.[422, 430]

Yet, the main take-away from *Westworld* is not that it answers these questions or predicts the future. Rather, the series shows us what our fears are in the present. *Westworld* sheds light on how we think of robots, at least in the eyes of the American film industry. We project onto robots our desire for power and assume they will behave in the same way if only given an opportunity to do so. And, when it relates to our own character, we do not trust ourselves to distinguish between good and evil.

If this is true, the challenge now is not a technological, but rather a moral and educational one. It is not robots that we should be afraid of, but ourselves. In turn, if the problem lies within us, so does the solution. We need to prepare for Relationships 5.0, in which we will interact with and even care for artificial beings, at least as we care for our pets and our environment, if not our colleagues and peers.

Even if some countries attempt to stem the evolution toward Relationships 5.0, other countries like China, Russia, and Japan will continue to press forward. They are already investing an incredible amount of money and human resources in developing such applications. It will only be a matter of time until such developments will be leaked to other countries, if not accepted wholeheartedly, as the world continues to globalize.

Perhaps we should take another approach instead of railing against current developments. We need to see how we can plan for the future in moral and just ways so we will adhere to appropriate standards of behavior.[431] We can prepare our children to live and act in the age of Society 5.0 and thoroughly discuss the values and laws that should guide us through the coming age. Educational programs dealing with these issues may be much more helpful here than fighting against technological advancement.

Furthermore, we should codify regulations on human-tech relationships that will frame and organize the various interactions between humans and artificial beings. The 2019 Eurobarometer survey, based on the responses of 32,543 people, aged 15 and older from 33 European countries, shows the real need for policy intervention in these areas. Respondents were asked what they believe to be the best way to ensure artificial intelligence's ethical development. 46.2% said they want a public policy intervention to ensure the ethical development of AI, while only 17.5% said the industry can regulate itself. The rest were unsure or gave no answer. As in my other studies, here too, younger generations were much more at ease with leaving those issues to the industry. The 15–24 age group, for example, held this view almost twice as much (24.4%) as the 55+ age group (13.3%), although, even among the young, most respondents preferred a policy intervention. Interestingly, men were more supportive of policy intervention (49.5%) than women (43.5%), although they also favor encouraging the development of emerging technologies more than women in my other studies.

Even though the many experts I interviewed for this book see no life-threatening risk in dealing with the developments described here, we at least need to interpret Elon Musk's warning that AI is probably "the biggest threat to our existence"[312] as a threat to our *moral* existence, and be careful in developing emotional technologies in a just way.

Governments, though, are still lagging. According to a 2020 report published by the research firm Cognilytica, no country had enacted specific legislation or regulation concerning the ethical use of emerging technologies or any bias issues in the application and development of AI, XR, and robots.[432] The main concerns that were addressed were the protection of online data, regulation of autonomous vehicles, and AI-powered weaponry. The EU, UK, Singapore, Australia, and a few other countries are all actively considering the ethical usage of emerging technologies and have advanced discussions around this topic, but they are still in a "wait and see" mode.

Yet we should anticipate ethical dilemmas and address them before we face these challenges in daily life. The answers we give will ease the public's fears and further support human-tech relationships in a benevolent way.

Therefore, although this book is not about the ethics of emerging technologies, I want to prompt the reader to think about the questions these technologies raise. One can think of three major types of problems we are facing in the age of Relationships 5.0: problems on "their" side, the in-between, and "our" side. The first type focuses on "them": how artificial beings are designed and whether there are biases encoded therein. The second type regards our interaction with artificial beings, such as privacy: how they keep our secrets. The third type is about us, humans, and what we go through when we make connections with artificial beings.

We do not necessarily need to ban technological progress, if that is even possible at all, but we do have the choice to prepare for such a reality by thinking about these problems. We can set guidelines and prepare solutions for these problems instead of shutting our eyes and letting the revolutions come. In this way, we will be ready for Relationships 5.0 much long before we find a series of robots on Walmart shelves.

Diversification of Artificial Beings

When we think about the population of artificial beings, a significant problem we might tackle is their diversity. Hence first on the list of Relationships 5.0's challenges would be gender equality. Numerous studies show that when designers or users assign a gender to a robot, stereotypes follow.[433, 434] In particular, female-like robots are perceived as warm and accommodating, while man-like robots are perceived as authoritative. One study tested participants' reactions to different images of robots that were adjusted to appear masculine or feminine, and human-like or machine-like. The study found that images of feminine robots, both machine-like and human-like, were ranked as more desirable for contact than masculine-appearing images. They elicited higher levels of warmth and perceived competence as well as lower levels of discomfort.[435] Another study followed individuals making morally problematic requests of male and female robots. The results showed that the male-faced robots were deemed less impolite than the female-faced robots when rejecting a command, while female robots seemed less authoritative.[436] Yet another study suggested that people subconsciously trust a male-voiced computer more than a female-voiced computer and view a dominant female voice as less friendly than a dominant male voice.[437]

Thus, the gendering of artificial beings should be used with caution as many of society's group dynamics and biases arguably get transferred into the design of robots. Currently, feminine robots are often promoted as secretaries and caretakers, jobs typically associated as female ones, whereas male-robots are often created for security jobs.[438] Creators want their products to sell, so exploiting certain biases will likely be to their advantage. For instance, if it is found that people are more drawn to females than males in a certain context, then creating a robot with a female appearance and voice will be more accepted than a male-robot. This situation does not merely reflect the existing stereotypes of our society, it also reinforces and validates certain biases. Exploiting such biases perpetuate inequality and stereotypes, and can go wildly unethical.[433]

This problem, however, is not without possible solutions. For example, NASA's robotic astronaut assistant, Robonaut, was designed to be gender-neutral. Yet, this did not go so well, and people kept referring to Robonaut in male terms. As Nicolaus Radford, a former NASA roboticist and one of Robonaut's lead engineers, explained: "99 out of 100 are quicker to identify a robot and use a 'he' pronoun. You know, 'Tell me what he can do!'"[439]

Similar attempts were made in the creation of a genderless voice called "Q." A product of several Danish organizations working collaboratively—Copenhagen Pride, Virtue, Equal AI, Koalition Interactive, and Thirtysoundsgood—Q aims to make new AI applications genderless. In its mission statement, its creators write: "As society continues to break down the gender binary, recognizing those who neither identify as male nor female, the technology we create should follow. Q is an example of what we hope the future holds; a future of ideas, inclusion, positions, and diverse representation in technology."[300] Yet again, many think this solution is challenging because, as Radford said, humans are usually drawn to a binary scheme and project onto artificial beings biases and stereotypes.

For this reason, the same Nicolaus Radford decided to explicitly challenge gender stereotypes. When he and his team designed a robot for stereotypical male jobs such as security and rescue in 2013, they hired a French graphic designer to help construct a woman-like robot. They called her Valkyrie, a tribute to the female warriors of Norse mythology.

This attempt was made despite NASA's official stance that robots should not have gender. NASA official statement was that Valkyrie's appearance originated from a need to move a 30-pound battery pack to its torso to balance its center of gravity. "Although it looks like it does have a female form, it's more a result of form and function versus actual design to be a female or a male," stated Jay Bolden, a NASA public affairs officer.

Radford, though, did not share NASA's official views on Valkyrie's gender. In an interview, he pointed to his 7-year-old-daughter, saying that Valkyrie is "a major source of inspiration for her. She talked about it all the time. She drew pictures of Valkyrie."[439, 440]

Other solutions include robots that are customizable, allowing a user to pick the robot's features for every part of its identity. For instance, a user could choose a typically "male" haircut for the robot, but a "female" voice, and have it act both politely and assertively. Such solutions also align with studies showing that users form a stronger alliance with adaptive robots and feel better around them.[441] Yet, in choosing this solution, we are trusting users not to force and reinforce current stereotypes. While the advantage of this solution is that companies, at least, will refrain from imposing such inequalities, some people may take on themselves to tilt their robots against social egalitarianism.

Alternatively, robots can be designed as non–human-like. Gender stereotypes are simply less relevant when a robot is designed to look like a teddy bear or is simply in an elliptical shape. However, this solution might still cause a problem since humans are usually drawn to specifically human-like objects or attribute human and gender qualities even without such signs.[442]

Alongside these solutions, some advocate for actively increasing the number of women among the designers and engineers of these technologies. Neil McArthur, from University of Manitoba, told me: "Most high-profile companies in this sphere are very male oriented. I think it's incumbent on technology companies to involve female designers and design products that are more female friendly. There is a lot of potential there."

Obviously, there are not perfect solutions, and I discuss some other, more general pathways to resolving this issue later in this chapter.

If this is not enough, other biases may also play out in Relationships 5.0. A team of researchers from New Zealand, Australia, China, and Germany found the race of artificial beings as an important factor. They measured American participants' reactions to dark-colored and light-colored robots, and found that racial biases extended to these robots. Participants assigned a race to the colored robots and reported relating negative attributions.[443]

Perhaps this is why the ethicist Robert Sparrow of Monash University in Australia found that most humanoid robots are either stylized with white material or have a bright metallic appearance. Manufacturers want to design products that sell, he argues, and because they find people to be racist, they prefer making white robots that are more likely to play into these biases.[444] In this way, similar to the lack of diversity in the media, the overrepresentation of white artificial beings reinforces a lack of racial diversity and perpetuates the underrepresentation of minorities. Moreover, it may also decrease trust and social acceptance amongst minorities who cannot find a robot similar to them, thereby preventing them from connecting to that robot on a deeper level.

Even if we increase the diversity of artificial beings, we should be careful with their assigned roles. If manufacturers start designing more racial minority robots, for example, there is no guarantee that they won't be primarily used as service robots instead of serving as socially equal robots. Not only would this reinforce the overrepresentation of minorities in domestic work, it would also leave the issue of minorities lacking artificial beings to identify with unsolved.[444]

Here, too, one potential solution to the issues surrounding race might be to create a "raceless" robot with unassociated colors. Another possible solution could be to exclusively increase social robots' racial diversity, but not *service* robots, keeping the race of service robots as primarily white or ambiguous. This might have the added benefit of challenging the overrepresentation of minorities in domestic fields and increasing racial equality.

Joining gender and racial dilemmas, another potential issue with respect to the diversity of artificial beings is their assigned age. Although exploring the ethics of age issues in artificial beings is critical, little attention has been paid to age diversification. Yet, most AI technology and robots that are currently in circulation have younger voices and appearances. On Replika, for example, you will not be able to find old characters. Most of the characters look to be in their 20s, if not younger. This threatens to solidify a stereotype that younger individuals are more capable, desirable, or knowledgeable than older individuals. Using the scenario of Denise and her social robot John, did anyone picture him as an old, wise man? Why not, as it is just as likely for someone like Denise to process her day at work with a peer or elder as with a younger assistant-type? Even to companies, it might well be fruitful to design elderly artificial beings and integrate them among different populations, young and old alike, to be both the source and subject of warmth, sympathy, and even admiration. Furthermore, artificial beings can be designed to grow old with us. If Denise chooses a robot of her own age as her first robot, for example, it might well be that John will grow old with her.

Similarly, no real investigation went into the exploration of the perceived class of a robot, its perceived nationality, or its perceived sexuality, both with respect to preventing prevailing stereotypes as well as to the possibility of using those robots to challenge existing stereotypes. For example, creating gay robots may help to promote the acceptance of homosexuality. These issues warrant future research and policy deliberation to make the advent of Relationships 5.0 as ethical as possible.

Because no solution is fault-proof, much effort should be invested in education. It is educational training, not technology, that should be developed. While artificial beings' diversification issues present a real challenge, they do not exist in isolation. These problems are a part of a broader social problem. If inequality exists in our society, then we are expected to see its unfortunate consequences among AI personal assistants, VR avatars, and robots. And vice versa, if we fight prejudices, emerging technologies will reflect this.

Therefore, we should educate future generations against the perils of gendering, racialization, and ethnicization in everything they do and interact with, including artificial beings. Moreover, beginning in elementary school, we should talk about the risks Relationships 5.0 entail as in any other interaction children have. We should let children experience the interaction with all sorts of artificial beings and guide them through. This attentive exposure should bear fruit later, when these children become adults and deal with all kinds of social dilemmas, among humans and robots alike.

Safe Connection

The diversification of AI personal assistants, virtual avatars, and robots is not the only problem present when exploring the ethicality of Relationships 5.0. In fact, it does not even scratch the surface. Another challenge comes in what relates to our connection with artificial beings, the in-between problem.

A major example of this type of problems is that of privacy. Human couples often expect that they can be vulnerable with each other and trust the other to keep embarrassing or delicate revelations private. Can an artificial companion be likewise trusted?

People may be very exposed and vulnerable with their robots and AI-based devices. A literature review that analyzed 60 studies on social robots found that sensitive information is commonly collected about users' everyday lives and their emotional and mental states. The consequences of such data being breached or even just mishandled can be harsh. For instance, children's casual comments about their personal feelings to their robots may lead their parents to spy on them against their will. In other cases, a worrying family may put their elderly relative into assisted care or limit their activities after accessing sensitive data that might be interpreted wrongly.[445]

This concern is especially acute in regards to Relationships 5.0. Current developments are focused on expanding the adaptivity of artificial companions, which means a focus on data collection.[441] With more data, AI-based robots autonomously adjust to users' preferences and become highly personalized. Such automatic personalization is shown to be a crucial capability for creating attachment. Once achieved to a high enough degree, personalization improves user engagement and allows more people to cohabit with robots.[446]

However, this personalization might also put privacy issues at risk. For example, users, like Denise, might share with the robot confidential information from work and trust it with their schedules and connections. Others may undress in front of their robots, thereby allowing the robot to record those private images. Indeed, one survey asked 449 users, ages 18–60, how they felt about the presence of connected devices in their lives. In response, users' main concern was nakedness (32%), identity theft (31%), and being overheard (26%). Interestingly, people were concerned about nakedness even in a scenario where the robot is not recording them. 74% of participants preferred a robot that turns around when they begin undressing.[447]

With XR, privacy concerns can be even more severe. XR devices can record both the content and nature of the interactions that users have in virtual worlds and such data can be breached and shared. In fact, users may be unaware of the extent of data collection of which the XR device is capable as the device seems only a mediator. The device may appear as just another pair of glasses. As a result, users may disclose more information than they otherwise would have felt comfortable with had they been aware of the XR device's full capabilities. This threat can be more severe in cases of more sophisticated XR devices that use BMI technology. Such devices can pick up on the brain waves of users and upload highly sensitive data to the cloud.

In addition, XR users often choose to share their content with other users via an extended reality app or Internet-based software. In some cases, the content of one user is shared with others without the original user's explicit consent. For example, users might be unaware that their view from a 3D camera might be shown to other users in real time if they are in their "team" or otherwise have given permission. This presents real problems for developers as they try to protect users' privacy and make the use of XR technologies more secure.[448]

With AI personal assistants, the problem is already widely discussed. While many users think of their AI assistants as merely entertainment or a way to turn on the light in the morning, their presence in our homes extends far beyond such voice commands. The open nature of the voice channel they use makes information leaking a real threat, and their constant connection to cloud computing exposes them to attacks.[449]

It is easy to see how a robot taken in for repairs could reveal private data and videos of its owners, how a stormy relationship with a digital avatar can be exposed by hacking the XR device, or how AI assistants record and transmit information that even close relatives should not have heard. However, as threatening as these scenarios sound, most of us already sit in front of a laptop with its camera directed at us, ready to record us at any time, sometimes without our knowledge.[450] Likewise, we carry our smartphones with us to the most private places and occasions without being really worried that the entire world, almost literally, is currently connected to that tiny device. Google, Meta, and Amazon already know about the connections we have, our property, and our shopping preferences. In short, privacy challenges with technology already exist and we learned to live with them. For most of us it is not a reason for sleepless nights out of worrisome.

Still, some solutions have already been offered in light of these concerns and could be integrated as part of future regulative measures. Several researchers, for example, designed a robot that can detect a user's nakedness based on imaging screenings and turn away if it detects that its user is changing, even before the user gets exposed.[447] Another team suggested increasing the level of transparency with artificial beings by allowing users to access, delete, or correct all audio and video data collected by them.[451] Still others suggested designing artificial beings that can signal when they are collecting data and that have an option to manually turn them off.[445]

Security and privacy solutions for extended reality are in the making as well. Technology is already being developed to integrate identity authentication into XR technologies. One possibility is the introduction of PIN codes and unlock patterns into XR headsets. A recent study tested this concept in laboratory conditions by giving 30 VR users the capability to create such codes or unlock patterns using a small handheld device that synchronizes with VR technology. The participants reported ease of use for both PIN codes and unlock patterns, with the average unlocking time being reduced to less than three seconds for both options after just one attempt. The developers note that there is little risk of outside observers surmising passwords through observing hand gesticulations. Given the need for security and the success of initial identity authentication processes, it seems likely that such measures will be integrated and upgraded with the development of XR.[452]

In parallel, a group of American researchers went to significant lengths in 2018 to establish guidelines for developing secure technologies that reduce the risk of privacy issues in XR. By surveying existing policies, conducting extensive interviews with XR users, and following new XR technologies, they published standards that all developers can use to help minimize the risk of privacy breaches. This *code of ethics*, as they call it, includes recommendations for transparency about information sharing in XR applications; social guidelines on how to avoid the creation of virtual spaces that put privacy or security at risk; and technical recommendations for practical requirements, such as reducing user nausea (a common problem in VR), that may otherwise compromise user autonomy. While these guidelines are in the early phase, they set a precedent and could serve as a foundation for the establishment of XR privacy laws in the future.[453]

We should, in any case, keep an eye on these issues and enact comprehensive and varied safeguards to protect us. Solutions should continue to be offered, and regulations put in place to make sure our usage is safe. Such an active approach is expected to pave the way for emerging technologies to become mainstream. Looking ahead, despite all risks and hurdles, it seems we will continue to balance privacy threats and our emotional and mental needs, exactly as we did when adopting previous technological developments.

Being Human in the Age of Relationships 5.0

Finally, if diversification of artificial beings is an example of concerns about the population of artificial beings, and privacy issues are instances where our interaction with them entails risks, we need to survey examples of ethical dilemmas that arise in regard to the changes we, human beings, will go through by having novel relationship opportunities in Society 5.0.

One such concern is that of objectification. Because AI assistants, robots, and digital avatars lie between a human and an object, some are afraid it may cause us to increasingly objectify other humans in our lives.[364] A full-scale friendship, for example, is not only instrumental in giving pleasure to the friends, but is inherently valuable, since friends will often support each other even when it comes at a cost to themselves. Human friendship causes the people involved to trust one another and be a source of support.[454] Yet, when we introduce an artificial being as a friend or romantic partner, we may threaten the inherently valuable part of a relationship. A robot may be a companion with which to spend a Sunday afternoon, because conversing with it is engaging and entertaining. It might also help its human partner feel more secure knowing that they have someone to talk to if they feel so inclined. But our relationships with the robot may not be inherently valuable in that we respect the robot for who it is. We would not be willing to make sacrifices for its benefit.

Moreover, while our potential relationships with artificial beings are one source of worry, many also fear that if humans increasingly turn to robots as substitutes for friends and romantic partners, they may start viewing other humans as they do robots: as an instrumental means to attain pleasure rather than an end onto themselves.[455]

If we turn to the prominent philosopher Martha Nussbaum to delve into the depths of human-to-human relationships, we can think of the risks of such objectification on several levels. Objectification, Nussbaum argues, can include treating others as if they lack autonomy, agency, boundaries, self-ownership, importance, and as if they can be exchanged with other objects.[454] When it comes to artificial beings these traits abound, and instincts we naturally possess with other humans when we treat them badly, such as guilt, can be missing. Similarly, virtues such as altruism and nobility can start to become uncommon in dealing with artificial beings. The real threat, therefore, is that we will start treating others as an extension to ourselves and as a tool to achieve greater ends.[456]

I heard these concerns time and again in the interviews I conducted. Natasha, a 33-year-old divorced woman from Washington, for example, stated:

I do not think romantic relationships with AI assistants is healthy or beneficial in any way. It stops people from connecting with actual people and could cause them to think it is okay to act however they want with people as AI assistants are unlikely to say no or turn them down, making people believe they are entitled to treat humans however they please as well.

Natasha, like many others, worries that by advancing tech companions, we threaten to promote cold, narcissistic relationships as the ideal to our future generations. We will reduce others to their bodies and appearance, treating them as if they lack the ability to speak independently.[457] Even worse, this may lead to an existential question we will ask ourselves: Are we really needed in a world in which a machine can replace our relationships, careers, and uniqueness?

Thus, the knock-on effects of integrating robots into our lives are extremely complicated and far-reaching. With the complexity of Relationships 5.0 comes the fear of the unknown that seems to only multiply our existing fears of everything that is out of the accepted norms.

Against these concerns, however, we can equally argue the opposite: what if human-tech relationships actually make us more caring and compassionate? A quick history lesson can show us how humanity has developed increasing compassion for decreasingly sentient beings, more caring of fellow humans, other species, and the environment as a whole. First, we began with developing empathy, morals, and compassion for other humans—that is, members of our own species—when our earliest human ancestors were innately programmed to care for one another in their own tribe.[56] These feelings slowly extended to other tribes, nations, and races, evolving to abolish and denounce slavery and letting morality prevail over human greed and injustice. Inequality and stratification are still abundant, but, looking back, we went a long way in this sense.

Humans have also increasingly engaged in cross-species empathy, showing care and concern for those who are different, as many other mammals instinctively do.[458] Historians and archaeologists agree that humans have long showed compassion for other species. Especially after humans have started to domesticate animals, they also began to enact laws against torturing them. Some animals, such as the dog and the cat, even became our closest friends. In fact, a recent excavation in Egypt has revealed the oldest known pet cemetery, dated to the first century AD.[459] The remains of 585 cats and dogs had been laid in prepared pits, some wearing collars and other adornments. By the nineteenth century, Queen Victoria had a dog called Dash that was widely known as her "closest childhood companion," and writers started to draw attention to animal welfare issues. Upton Sinclair, for example, lamented animal suffering in his 1906 book, *The Jungle*. It was not long before governments began to legislate against animal torture.[460]

Furthermore, recent history has taught us that human compassion is not limited to other humans or only the sentient animals that we know and care for personally. The accelerating popularity of vegetarianism and veganism is fueled by compassion for animals that we have not personally known or domesticated. In this way, the development of animal welfare extended beyond laws against violence or the closing of slaughterhouses.[461]

To go even one step further, the growing popularity of environmentalism and concerns about climate change show that humans are worried now about the welfare of the inanimate. In other words, humans have become more and more inclusive of and compassionate toward beings that are less and less human-like.

Thus, it seems that throughout human history, despite all headwinds, society only became more inclusive of its surroundings. In the same way, we could very well see parallel processes with Relationships 5.0. There is a possibility that instead of being dehumanized as technology gains "life," we will actually extend our compassion further to objects around us. As shown in this book, we are surprisingly capable of developing feelings toward technology, so it will not be such a stretch if we learn to respect electronic creations. Instead of objectifying our close human friends and partners, we will learn to feel gratitude and respect toward technology. As new technological developments become more integrated into our lives, more people will develop empathy for their digital partners. In this sense, the three revolutions described here—the cognitive, sensorial, and physical—together elevate the inanimate. They demonstrate how even objects can and should be respected. As strange as it may seem, especially to Western readers, it might well be the main moral lesson we need to learn from Relationships 5.0.

Societies such as Japan and South Korea, where the integration of emerging technologies is already prevalent, reflect these processes. In these cultures, robots are seen as deserving of warmth, sensitivity, and compassion.[462] According to Jennifer Robertson, professor of Anthropology and the History of Art at the University of Michigan, the Japanese approach toward technological creations is based on the concept of Shinto, the animistic beliefs about life and death.[26, 463] Unlike the three major monotheist religions, Shinto holds that energetic forces called kami exist in all aspects of the world and universe. Kami infuses humans but also animals, insects, trees, rocks, and human creations, like robots or virtual artifacts. Such attitudes can quickly become universal, especially in a growingly secular world, and make common ground in understanding our relationships with technology.

We already hear calls to discuss the legal and moral rights of robots and other technologies.[464] Given the speed at which ideas travel in our globalized world, it is not too farfetched to suggest that these calls will eventually—and quite quickly—turn into social movements with legal demands. Exactly as we respect living beings around us and become attached to them, it might well be that we will set rules and guidelines for how to treat AI assistants, digital avatars, and robots.[456]

Therefore, in considering the moral and ethical questions that will emerge from Relationships 5.0, it seems that the answers to future challenges may be reduced to the question of whether we believe in the goodness of humanity. If we trust humanity, we will be able to prepare for the oncoming changes by integrating morals and ethics into the era of Relationships 5.0. How we choose to develop and treat new technologies will be reflected in the ways that AI, virtual beings, and robots integrate with humans. If we choose compassion, the dark scenarios of *Westworld* will remain fictional.

In any case, putting these processes on a historical continuum and thinking about their character today urge us to make our call. If the increase in human-tech rapprochement makes sense and has become almost deterministic with recent technological developments, we should start being serious about it. We can oppose the coming changes, but it might be more helpful to adjust ourselves accordingly.

It will be hard to make sure we take the right turns on our road there, but it seems feasible. We should apply all social developments we acquired over the years, such as egalitarianism and social justice, to Relationships 5.0. Guided exposure to technological developments that already surround us, such as AI personal assistants, and thoughtful preparation for a future of widespread robotics and virtual worlds are required now. Engineers, designers, and tech entrepreneurs should assume more responsibility in making race-, gender-, and age-sensitive products; and, educators, spiritual leaders, and moral influencers should prepare society for a stage where we respect our technological creations. Ethics should be developed and enforced, and regulations should be in place to prevent biases and stereotypes.

If we do so on time, we can create a reality in which technologies achieve the exact opposite of the *Westworld* dystopian reality, wherein social inequalities and biases are amplified by Relationships 5.0. By carefully preparing ourselves, we can create a future where connections strengthen, new friendships are formed, and emotional attachments become varied, among humans and artificial beings alike. Moreover, if AI, VR, and robots are designed and programmed to promote equality and equity, we can see a future in which we flourish.

Conclusion

When I started writing this book, I mainly thought about the technological revolutions we are gearing up to experience: will artificial intelligence, virtual reality, and robotics affect our emotional lives? By now, however, I feel I must go back to a much more basic question: What is the nature of that special connection many of us have, or wish to have, with a significant other? Or, more simply: What is the meaning of love?

We are all exposed to representations of finding a soulmate, falling in love, and living happily ever after. By definition, a "soulmate" cannot be scientifically analyzed and certainly does not belong to the realm of technology and machines, but to the realm of the "soul"— no matter what that means. Falling in love seems almost a religious term. It is a matter of faith and fate, rather than being the actual building blocks that address our needs.

We still strongly believe that only humans can trigger real feelings in us. "Love cannot be engineered," I heard time and again in the interviews I conducted for this book. It seems so true: I think of the flare of emotions we encounter when falling in love, the aching feeling when recounting painful experiences with a close friend, or the cascade of relief when discovering that a loved one's chemotherapy was successful. How awkward and distasteful it would be to experience these moments and emotions with a machine. Can robots emulate that eruptive catharsis of liberation and gratitude when we offload our most private secrets? Can conversations with an AI help us heal emotional scars, coating them with understanding and affection?

We still live in an age, reminiscent of previous eras, when we are convinced that nothing can be designed to replicate these feelings. Only sharing our emotions with other humans makes us feel comforted as we reveal layers of our personality and smooth out the wrinkles of the past. Current technologies do not seem to be able to meet our emotional needs, and we struggle to imagine this possibility in the future.

Yet, if we allow ourselves to suspend these notions, even just for one moment, we can see that love, marriage, and bonding are less "magical" than we at first thought. Looking back at evolutionary processes and the history of humankind, as detailed in Part I of this book, showcases how love and relationships were much less emotional and idealized than we think of them today. In fact, the emotions we feel today can be thought of as the collection of instincts we gather throughout our long history.

At the point when we first fall in love or get excited about someone, we are basically excited about a collection of signals we receive. This collection of subtle characteristics, nuanced messages, and elusive hopes are based on our needs to be fed, procreate, and be guarded. Perhaps it is how the other person smells—which some claim to be connected to evolutionary senses—that tells us who will be best to procreate with.[465] In other cases, it can be the way they touch us—soft enough to make us feel relaxed and nurtured or maybe strong and steady to make us feel secure and safe. To complete this set of signals, we might also focus on their smile, which makes us sense how sweet they are at heart and how able to take care of us in rough times, though whether this is true is something we only learn later, in the ups and downs we face together. As studies have shown, the expectation that this person will satisfy our needs releases neurotransmitters and hormones like dopamine and oxytocin that give us this sensation of "love." Through this release, which can be artificially produced or oppressed quite easily, our neural systems learn to want more time with a particular partner.[466] Today, however, it still translates into abstract emotions that make us feel only humans can make this happen properly.

Thinking in this way, as cynical or overly scientific as it might sound, helps overcoming preconceptions of "proper love" and start recognizing and embracing what technology can offer. We can insist on "naturally grown love," but we can also start thinking about the possibility of consummating affection, care, and empathy through technology, even if only partly at first.

As shown throughout Part II, acts of love and warmth, produced responsibly by technology, have already been found to create responses similar to those we get from other humans. Technology has increasingly shown the power to satisfy some of our most sophisticated and nuanced emotional needs. Many components of human desires that we used to think of as irreplaceable, almost mystic, can be produced by technology now or will be in the near future.

In this way, the three technological revolutions described in this book are paving a way to something fundamentally different. The cognitive, sensorial, and physical revolutions are likely to change civilization in the same way that fire, agriculture, and steam engines did when they were discovered and invented by humans. Technological advancements have opened the door to the convergence of humans and technology: Vincent, the cute chatbot that embarrassingly admits his faults, the virtual bartender who seduces people in a virtual bar and makes them feel guilty, and Erica the actroid who looks and behaves like a real Japanese woman are all part of our world today. Especially following the COVID-19 pandemic, today's society is accelerating the adoption of new forms of technology. Applying them to relationships will not be an exception.[467]

Naturally, these changes come with both advantages and disadvantages. Readers who are deeply unsettled by the contents of this book, as well as those who are highly enthused by the prospect of human-robot relationships, must come to the same agreement: what characterizes our species and has kept us in existence throughout history is our ability to evolve and adapt to new conditions. We should not be surprised, therefore, that humanity is adapting to technology once again.

Regardless of our normative stance, we have a responsibility to explore, anticipate, and research how these technologies will affect human behavior, emotion, and society. Without this kind of research, we will be dismally ill-prepared to overcome the obstacles and challenges that will arise from the inevitable changes to the fabric of human society and personal relationships. Hopefully, this book moved us one step further toward this goal.

Acknowledgments

I cannot express enough thanks to the many people who saw me through this book. I am especially grateful to those surrounding me and close to me, whose love and appreciation are invaluable. Their unwavering support and encouragement pushed me to write and finish this book.

This book is certainly not mine alone. I worked on this book with countless brilliant and diligent people, and I owe them my deep gratitude. I wish to thank all the research assistants, proofreaders, interviewees, and experts, who assisted me immensely in making this book what it is. Above all, I would like to thank my agent, Max Sinsheimer, whose help throughout the process was truly remarkable. I would also like to thank the extraordinary team of people at Oxford University Press, whose professionalism I admire. Last but certainly not least, I thank my dear editor, Sarah Humphreville, for believing in this book and providing me with great support, insights, and comments along the way. Thank you all.

References

Introduction

1 Benjamin Haas, "Chinese Man 'Marries' Robot He Built Himself," *The Guardian* (2017); Claire Heffron and Tracy You, "Engineer, 31, Who Married His Robotic Girlfriend to Appease His Pushy Parents Is Now Making Customised Sex Robots for Millions of Bachelors" (2017), https://www.dailymail.co.uk/news/china/article-5021113/Engineer-31-marries-inflatable-robot-girlfriend.html.

2 Parmy Olson, "My Girlfriend Is a Chatbot" (2020), https://www.wsj.com/articles/my-girlfriend-is-a-chatbot-11586523208.

3 Shuaishuai Wang, "Calculating Dating Goals: Data Gaming and Algorithmic Sociality on Blued, a Chinese Gay Dating App," *Information, Communication & Society*, 23 (2020), 181–97; Daniel B. Shank, Christopher Graves, Alexander Gott, Patrick Gamez, and Sophia Rodriguez, "Feeling our Way to Machine Minds: People's Emotions When Perceiving Mind in Artificial Intelligence," *Computers in Human Behavior*, 98 (2019), 256–66.

4 Elisabeth Timmermans and Cédric Courtois, "From Swiping to Casual Sex and/or Committed Relationships: Exploring the Experiences of Tinder Users," *The Information Society*, 34 (2018), 59–70.

5 Aziz Ansari and Eric Klinenberg, *Romance Moderne* (New York: Penguin, 2017).

6 Tinder, "About Tinder," Tinder (2019), https://www.gotinder.com/press.

7 Yuval Noah Harari, *Sapiens: A Brief History of Humankind* (New York: Random House, 2014).

8 Gerhard Emmanuel Lenski and Jean Lenski, *Human Societies: An Introduction to Macrosociology* (McGraw-Hill Companies, 1982); Elman R. Service, *Origins of the State and Civilization* (New York: Norton, 1975).

9 Yoshihiro Shiroishi, Kunio Uchiyama, and Norihiro Suzuki, "Society 5.0: For Human Security and Well-Being," *Computer*, 51 (2018), 91–95.

10 Klaus Schwab, *The Fourth Industrial Revolution* (New York: Crown Business, 2017).

11 Toby Stuart and Chris Anderson, "3d Robotics: Disrupting the Drone Market," *California Management Review*, 57 (2015), 91–112; Chin-Ching Yeh, "Trend Analysis for the Market and Application Development of 3d Printing," *International Journal of Automation and Smart Technology*, 4 (2014), 1–3.

12 Jianfeng Gao, Michel Galley, and Lihong Li, "Neural Approaches to Conversational AI," *Foundations and Trends® in Information Retrieval*, 13 (2019), 127–298.

13 Jiahuan Pei, Pengjie Ren, and Maarten de Rijke, "A Modular Task-Oriented Dialogue System Using a Neural Mixture-of-Experts," *arXiv preprint arXiv:1907.05346* (2019); Mikhail Burtsev, Alexander Seliverstov, Rafael AIrapetyan, Mikhail Arkhipov, Dilyara Baymurzina, Nickolay Bushkov, Olga Gureenkova, Taras Khakhulin, Yuri Kuratov, and Denis Kuznetsov, "Deeppavlov: Open-Source Library for Dialogue Systems," in *Proceedings of ACL 2018, System Demonstrations* (2018), 122–27.

14 Business-Insider, "Jeff Bezos Says the Echo 'Isn't About' Getting People to Shop on Amazon, and He May Be Right" (2017).

15 Kathleen Kara Fitzpatrick, Alison Darcy, and Molly Vierhile, "Delivering Cognitive Behavior Therapy to Young Adults with Symptoms of Depression and Anxiety Using a Fully Automated Conversational Agent (Woebot): A Randomized Controlled Trial," *JMIR Mental Health*, 4 (2017), e19.

16 IDC, "Augmented Reality and Virtual Reality Headsets Poised for Significant Growth" (Framingham, MA: International Data Corporation, 2020).

17 Benedict Anderson, *Imagined Communities: Reflections on the Origin and Spread of Nationalism* (London and New York: Verso Books, 1983).

18 Alex Heath, "The People with Power at Facebook as its Hardware, Commerce Ambitions Expand" (The Information, 2021).

19 Kenneth Research, "Global Humanoid Robots Market by Component, Product, Application and Region 2014–2025" (New York: Kenneth Research, 2019), p. 222; IndustryARC, "Humanoid Robot Market—Forecast: 2020–2025" (Hyderabad, Telangana: IndustryARC, 2019), p. 222.

20 "Boston Dynamics COVID-19 Response," Boston Dynamics (2020), https://www.bostondynamics.com/COVID-19; Mohammad Nasajpour, Seyedamin Pouriyeh, Reza M. Parizi, Mohsen Dorodchi, Maria Valero, and Hamid R. Arabnia, "Internet of Things for Current COVID-19 and Future Pandemics: An Exploratory Study," *arXiv preprint arXiv:2007.11147* (2020); Evan Ackerman, "Autonomous Robots Are Helping Kill Coronavirus in Hospitals," *IEEE Spectrum* (2020).

21 Michał Dziergwa, Mirela Kaczmarek, Paweł Kaczmarek, Jan Kędzierski, and Karolina Wadas-Szydłowska, "Long-Term Cohabitation with a Social Robot: A Case Study of the Influence of Human Attachment Patterns," *International Journal of Social Robotics*, 10 (2018), 163–76.

22 Nick Bostrom, "The Superintelligent Will: Motivation and Instrumental Rationality in Advanced Artificial Agents," *Minds and Machines*, 22 (2012), 71–85.

23 Alexandre Coninx, Paul Baxter, Elettra Oleari, Sara Bellini, Bert Bierman, O. Henkemans, Lola Cañamero, Piero Cosi, Valentin Enescu, and R. Espinoza, "Towards Long-Term Social Child-Robot Interaction: Using Multi-Activity Switching to Engage Young Users," *Journal of Human-Robot Interaction*, 5 (2016).

24 Rhea Cheema, "Sex Doll Robot Samantha Can Now Say No to Unwanted Sexual Advances," *India.com* (2018).

25 Nicola Döring, M. Rohangis Mohseni, and Roberto Walter, "Design, Use, and Effects of Sex Dolls and Sex Robots: Scoping Review," *Journal of Medical Internet Research*, 22 (2020), e18551.

26 Jennifer Robertson, "Robo Sapiens Japanicus: Humanoid Robots and the Posthuman Family," *Critical Asian Studies*, 39 (2007), 369–98.

27 Government of Japan, "Innovation 25 Interim Report" (Tokyo, Japan: Prime Minister of Japan and his Cabinet, 2007); Jennifer Robertson, "Human Rights vs. Robot Rights: Forecasts from Japan," *Critical Asian Studies*, 46 (2014), 571–98.

28 Shimon Ullman, "Using Neuroscience to Develop Artificial Intelligence," *Science*, 363 (2019), 692–93.

29 Demis Hassabis, Dharshan Kumaran, Christopher Summerfield, and Matthew Botvinick, "Neuroscience-Inspired Artificial Intelligence," *Neuron*, 95 (2017), 245–58.

30 Chris Mack, "The Multiple Lives of Moore's Law," *IEEE Spectrum*, 52 (2015), 31–31.

31 Ray Kurzweil, *The Singularity Is Near: When Humans Transcend Biology* (New York: Penguin, 2005).

32 David J. D. Earn, Junling Ma, Hendrik Poinar, Jonathan Dushoff, and Benjamin M. Bolker, "Acceleration of Plague Outbreaks in the Second Pandemic," *Proceedings of the National Academy of Sciences*, 117 (2020), 27703–11.

33 Christopher L. Magee and Tessaleno C. Devezas, "How Many Singularities Are Near and How Will They Disrupt Human History?," *Technological Forecasting and Social Change*, 78 (2011), 1365–78.

34 Hua Wang and Barry Wellman, "Social Connectivity in America: Changes in Adult Friendship Network Size from 2002 to 2007," *American Behavioral Scientist*, 53 (2010), 1148–69.

35 Patrick Van-Kessel, *How Americans Feel about the Satisfactions and Stresses of Modern Life* (Washington, DC: Pew Research Center, Social & Demographic Trends Project (2020).

36 "Each of the 28 Eu Member States as Well as from the Five Candidate Countries (Turkey, North Macedonia, Montenegro, Serbia, and Albania) and the Turkish Cypriot Community."

Chapter 1

37 Lawrence Barham, *From Hand to Handle: The First Industrial Revolution* (Oxford: Oxford University Press, 2013).

38 Richard W. Wrangham, James Holland Jones, Greg Laden, David Pilbeam, NancyLou Conklin-Brittain, C. Loring Brace, Henry T. Bunn, Eudald Carbonell Roura, K. Hawkes, and J. F. O'Connell, "The Raw and the Stolen: Cooking and the Ecology of Human Origins," *Current Anthropology*, 40 (1999), 567–94;

Manuel Domínguez-Rodrigo, Travis Rayne Pickering, Sileshi Semaw, and Michael J. Rogers, "Cutmarked Bones from Pliocene Archaeological Sites at Gona, Afar, Ethiopia: Implications for the Function of the World's Oldest Stone Tools," *Journal of Human Evolution*, 48 (2005), 109–21.

39 Katherine D. Zink and Daniel E. Lieberman, "Impact of Meat and Lower Palaeolithic Food Processing Techniques on Chewing in Humans," *Nature*, 531 (2016), 500.

40 Efraim Lev, Mordechai E. Kislev, and Ofer Bar-Yosef, "Mousterian Vegetal Food in Kebara Cave, Mt. Carmel," *Journal of Archaeological Science*, 32 (2005), 475–84.

41 Alain Testart, Richard G. Forbis, Brian Hayden, Tim Ingold, Stephen M. Perlman, David L. Pokotylo, Peter Rowley-Conwy, and David E. Stuart, "The Significance of Food Storage among Hunter-Gatherers: Residence Patterns, Population Densities, and Social Inequalities [and Comments and Reply]," *Current Anthropology*, 23 (1982), 523–37.

42 Susan Bulmer and Ralph Bulmer, "The Prehistory of the Australian New Guinea Highlands," *American Anthropologist*, 66 (1964), 39–76; Maxime Aubert, Rustan Lebe, Adhi Agus Oktaviana, Muhammad Tang, Basran Burhan, Andi Jusdi, Budianto Hakim, Jian-xin Zhao, I. Made Geria, and Priyatno Hadi Sulistyarto, "Earliest Hunting Scene in Prehistoric Art," *Nature*, 576 (2019), 442–45.

43 John D. Speth and Eitan Tchernov, "Neandertal Hunting and Meat-Processing in the Near East," *Meat-Eating and Human Evolution* (2001), 52–72.

44 Peter S. Ungar, *Evolution of the Human Diet: The Known, the Unknown, and the Unknowable* (Oxford: Oxford University Press, 2006).

45 Laure Dubreuil, "Long-Term Trends in Natufian Subsistence: A Use-Wear Analysis of Ground Stone Tools," *Journal of Archaeological Science*, 31 (2004), 1613–29; Ran Barkai, Cristina Lemorini, and Avi Gopher, "Palaeolithic Cutlery 400,000–200,000 Years Ago: Tiny Meat-Cutting Tools from Qesem Cave, Israel," *Antiquity*, 84 (2010), Online.

46 Zhaoyu Zhu, Robin Dennell, Weiwen Huang, Yi Wu, Shifan Qiu, Shixia Yang, Zhiguo Rao, Yamei Hou, Jiubing Xie, and Jiangwei Han, "Hominin Occupation of the Chinese Loess Plateau Since about 2.1 Million Years Ago," *Nature*, 559 (2018), 608.

47 Jared M. Diamond, *Guns, Germs and Steel: A Short History of Everybody for the Last 13,000 Years* (New York: W. W. Norton & Company, 1998).

48 SM Cooper, "Optimal Hunting Group Size: The Need for Lions to Defend their Kills against Loss to Spotted Hyaenas," *African Journal of Ecology*, 29 (1991), 130–36.

49 Claude Masset, "Prehistory of the Family," in *A History of the Family*, ed. Andre Burguiere, Christiane Klapisch-Zuber, Martine Segalen, and Francois Zonabend (Cambridge: Polity Press, 1996), 71–91.

50 Dani Nadel, Ehud Weiss, Orit Simchoni, Alexander Tsatskin, Avinoam Danin, and Mordechai Kislev, "Stone Age Hut in Israel Yields World's Oldest Evidence of Bedding," *Proceedings of the National Academy of Sciences*, 101 (2004),

6821–26; Dani Nadel and Ella Werker, "The Oldest Ever Brush Hut Plant Remains from Ohalo Ii, Jordan Valley, Israel (19,000 Bp)," *Antiquity*, 73 (1999), 755–64.

51 Lewis R. Binford, "Mobility, Housing, and Environment: A Comparative Study," *Journal of Anthropological Research*, 46 (1990), 119–52; Richard Bradley, *The Past in Prehistoric Societies* (New York: Routledge, 2014).

52 Nena Galanidou, "Patterns in Caves: Foragers, Horticulturists, and the Use of Space," *Journal of Anthropological Archaeology*, 19 (2000), 243–75; Lyn Wadley, "The Use of Space in the Late Middle Stone Age of Rose Cottage Cave, South Africa," in *Transitions before the Transition* (New York: Springer, 2006), 279–94.

53 Eric Alden Smith, Kim Hill, Frank W. Marlowe, David Nolin, Polly Wiessner, Michael Gurven, Samuel Bowles, Monique Borgerhoff Mulder, Tom Hertz, and Adrian Bell, "Wealth Transmission and Inequality among Hunter-Gatherers," *Current Anthropology*, 51 (2010), 19–34.

54 James Suzman, *Work: A Deep History, from the Stone Age to the Age of Robots* (New York: Penguin, 2021).

55 Karen L. Endicott, "Gender Relations in Hunter-Gatherer Societies," in *The Cambridge Encyclopedia of Hunters and Gatherers*, ed. Richard Lee and Richard Daly (Cambridge: Cambridge University Press, 1999), 411–18.

56 Kim Hill, "Altruistic Cooperation during Foraging by the Ache, and the Evolved Human Predisposition to Cooperate," *Human Nature*, 13 (2002), 105–28.

57 Dean J. Saitta, "Marxism, Prehistory, and Primitive Communism," *Rethinking Marxism*, 1 (1988), 145–68.

58 David Erdal and Andrew Whiten, "Egalitarianism and Machiavellian Intelligence in Human Evolution," in *Modelling the Early Human Mind*, ed. Paul Mellars and Kathleen Gibson (Cambridge: Cambridge MacDonald Monograph Series, 1996), 139–50; Adrienne L. Zihlman, "Woman the Gatherer," in *Reader in Gender Archaeology*, ed. Kelley Hays-Gilpin and David S. Whitle (New York: Routledge, 1981).

59 David Erdal, Andrew Whiten, Christopher Boehm, and Bruce Knauft, "On Human Egalitarianism: An Evolutionary Product of Machiavellian Status Escalation?" *Current Anthropology*, 35 (1994), 175–83; Christopher Boehm and Christopher Boehm, *Hierarchy in the Forest: The Evolution of Egalitarian Behavior* (Cambridge, MA: Harvard University Press, 2009).

60 T. Paoli, Elisabetta Palagi, and S. M. Borgognini Tarli, "Reevaluation of Dominance Hierarchy in Bonobos (Pan Paniscus)," *American Journal of Physical Anthropology: The Official Publication of the American Association of Physical Anthropologists*, 130 (2006), 116–22.

61 Frans B. M. Waal, *Our Inner Ape: A Leading Primatologist Explains Why We Are Who We Are* (New York: Riverhead Books, 2005).

62 Brian Hare, Alicia P. Melis, Vanessa Woods, Sara Hastings, and Richard Wrangham, "Tolerance Allows Bonobos to Outperform Chimpanzees on a Cooperative Task," *Current Biology*, 17 (2007), 619–23.

63 Michael Gurven, "To Give and to Give Not: The Behavioral Ecology of Human Food Transfers," *Behavioral and Brain Sciences*, 27 (2004), 543–59.

64 Kim R. Hill, Robert S. Walker, Miran Božičević, James Eder, Thomas Headland, Barry Hewlett, A. Magdalena Hurtado, Frank Marlowe, Polly Wiessner, and Brian Wood, "Co-Residence Patterns in Hunter-Gatherer Societies Show Unique Human Social Structure," *Science*, 331 (2011), 1286–89.

65 Barry S. Hewlett, *Hunter-Gatherer Childhoods: Evolutionary, Developmental, and Cultural Perspectives* (New York: Routledge, 2017).

66 Coren L. Apicella and Alyssa N. Crittenden, "Hunter-Gatherer Families and Parenting," in *The Handbook of Evolutionary Psychology*, ed. David Buss (New York: John Wiley & Sons, 2015), 1–20.

67 Nicholas G. Blurton Jones, Frank W. Marlowe, Kristen Hawkes, and James F. O'Connell, "Paternal Investment and Hunter-Gatherer Divorce Rates," in *Adaptation and Human Behavior* (New York: Routledge, 2017), 69–90.

68 Richard Borshay Lee and Irven DeVore, *Man the Hunter* (New York: Routledge, 2017).

69 Steven Mithen, "Simulating Prehistoric Hunter-Gatherer Societies," in *Simulating Societies* (New York: Routledge, 2018), 165–93.

70 Barbara Bender, "Gatherer-Hunter to Farmer: A Social Perspective," *World Archaeology*, 10 (1978), 204–22.

71 Sewall Wright, *The Effects of Inbreeding and Crossbreeding on Guinea Pigs* (Washington, DC: US Government Printing Office, 1922).

72 J. W. Jamieson, "Biological Diversity and Ethnic Identity: Changing Patterns in the Modern World," *Mankind Quarterly*, 36 (1995), 193.

73 Anne E. Pusey, "Inbreeding Avoidance in Chimpanzees," *Animal Behaviour*, 28 (1980), 543–52.

74 Erica M. Tennenhouse, "Inbreeding Avoidance in Male Primates: A Response to Female Mate Choice?," *Ethology*, 120 (2014), 111–19.

75 Susan Pfeiffer, L. Elizabeth Doyle, Helen K. Kurki, Lesley Harrington, Jaime K. Ginter, and Catherine E. Merritt, "Discernment of Mortality Risk Associated with Childbirth in Archaeologically Derived Forager Skeletons," *International Journal of Paleopathology*, 7 (2014), 15–24.

76 Kim Hill, A. Magdalena Hurtado, and Robert S. Walker, "High Adult Mortality among Hiwi Hunter-Gatherers: Implications for Human Evolution," *Journal of Human Evolution*, 52 (2007), 443–54; Michael Gurven and Hillard Kaplan, "Longevity among Hunter-Gatherers: A Cross-Cultural Examination," *Population and Development Review*, 33 (2007), 321–65.

77 William T. Divale, "Systemic Population Control in the Middle and Upper Palaeolithic: Inferences Based on Contemporary Hunter-Gatherers," *World Archaeology*, 4 (1972), 222–43.

78 Brian Hayden, "Population Control among Hunter/Gatherers," *World Archaeology*, 4 (1972), 205–21.

79 Frans B. M. De Waal and Sergey Gavrilets, "Monogamy with a Purpose," *Proceedings of the National Academy of Sciences*, 110 (2013), 15167–68.

80 William Engelbrecht, "Factors Maintaining Low Population Density among the Prehistoric New York Iroquois," *American Antiquity*, 52 (1987), 13–27.

81 Anne Innis Dagg, "Infanticide by Male Lions Hypothesis: A Fallacy Influencing Research into Human Behavior," *American Anthropologist*, 100 (1998), 940–50; Craig Packer and Anne E. Pusey, "Adaptations of Female Lions to Infanticide by Incoming Males," *The American Naturalist*, 121 (1983), 716–28.

82 William Tucker, *Marriage and Civilization: How Monogamy Made Us Human* (Washington, DC: Regnery Publishing, 2014).

83 Nicole M. Waguespack, "The Organization of Male and Female Labor in Foraging Societies: Implications for Early Paleoindian Archaeology," *American Anthropologist*, 107 (2005), 666–76.

84 Maren Huck, Anthony Di Fiore, and Eduardo Fernandez-Duque, "Of Apples and Oranges? The Evolution of "Monogamy" in Non-Human Primates," *Frontiers in Ecology and Evolution*, 7 (2019), 472.

85 C. Owen Lovejoy, "Reexamining Human Origins in Light of Ardipithecus Ramidus," *Science*, 326 (2009), 74–74e8.

86 Dieter Lukas and Timothy Hugh Clutton-Brock, "The Evolution of Social Monogamy in Mammals," *Science*, 341 (2013), 526–30.

87 Christopher Opie, Quentin D. Atkinson, Robin I. M. Dunbar, and Susanne Shultz, "Male Infanticide Leads to Social Monogamy in Primates," *Proceedings of the National Academy of Sciences*, 110 (2013), 13328–32.

88 Alan R. Rogers and Arindam Mukherjee, "Quantitative Genetics of Sexual Dimorphism in Human Body Size," *Evolution*, 46 (1992), 226–34.

89 Kim Hill and A. Magdalena Hurtado, "Cooperative Breeding in South American Hunter–Gatherers," *Proceedings of the Royal Society B: Biological Sciences*, 276 (2009), 3863–70.

90 Karen L. Kramer and Russell D. Greaves, "Postmarital Residence and Bilateral Kin Associations among Hunter-Gatherers," *Human Nature*, 22 (2011), 41.

91 Karen L. Kramer, Ryan Schacht, and Adrian Bell, "Adult Sex Ratios and Partner Scarcity among Hunter–Gatherers: Implications for Dispersal Patterns and the Evolution of Human Sociality," *Philosophical Transactions of the Royal Society B: Biological Sciences*, 372 (2017), 20160316.

92 Devra G. Kleiman, "Monogamy in Mammals," *The Quarterly Review of Biology*, 52 (1977), 39–69.

93 Lewis Henry Morgan, *Ancient Society: Researches in the Lines of Human Progress from Savagery, through Barbarism to Civilization* (New York: Henry Holt, 1877).

94 Christopher Ryan and Cacilda Jethá, *Sex at Dawn: The Prehistoric Origins of Modern Sexuality* (New York: Harper, 2010).

95 Mark Dyble, Gul Deniz Salali, Nikhil Chaudhary, A. Page, D. Smith, J. Thompson, Lucio Vinicius, Ruth Mace, and A. B. Migliano, "Sex Equality Can Explain the Unique Social Structure of Hunter-Gatherer Bands," *Science*, 348 (2015), 796–98.

96 Katherine E. Starkweather and Raymond Hames, "A Survey of Non-Classical Polyandry," *Human Nature*, 23 (2012), 149–72.

97 Robert S. Walker, Mark V. Flinn, and Kim R. Hill, "Evolutionary History of Partible Paternity in Lowland South America," *Proceedings of the National Academy of Sciences*, 107 (2010), 19195–200.

98 Isabelle Dupanloup, Luísa Pereira, Giorgio Bertorelle, Francesc Calafell, Maria Joao Prata, Antonio Amorim, and Guido Barbujani, "A Recent Shift from Polygyny to Monogamy in Humans Is Suggested by the Analysis of Worldwide Y-Chromosome Diversity," *Journal of Molecular Evolution*, 57 (2003), 85–97.

Chapter 2

99 O. Bar-Yosef, "Chapter 19—Multiple Origins of Agriculture in Eurasia and Africa," in *On Human Nature*, ed. Michel Tibayrenc and Francisco J. Ayala (San Diego: Academic Press, 2017), 297–331.

100 Marshall Sahlins, *Stone Age Economics* (New York: Routledge, 2017).

101 Ofer Bar-Yosef, "The Upper Paleolithic Revolution," *Annual Review of Anthropology*, 31 (2002), 363–93.

102 T. Douglas Price and Ofer Bar-Yosef, "The Origins of Agriculture: New Data, New Ideas: An Introduction to Supplement 4," *Current Anthropology*, 52 (2011), S163–S74.

103 John Burnett, *A Social History of Housing, 1815–1985* (London: Methuen, 1986).

104 Christopher Watson, "Trends in World Urbanisation," in *Proceedings of the 1st International Conference on Urban Pests, BPCC Wheatons Ltd, Exeter* (1993), 1–7.

105 Norman Yoffee, *The Cambridge World History: Volume 3, Early Cities in Comparative Perspective, 4000 BCE–1200 CE* (Cambridge: Cambridge University Press, 2015).

106 Mary Louise Flint and Robert van den Bosch, "A History of Pest Control," in *Introduction to Integrated Pest Management* (New York: Springer, 1981), 51–81.

107 Marek Zvelebil, "The Agricultural Transition and the Origins of Neolithic Society in Europe," *Documenta Praehistorica*, 28 (2001), 1–26.

108 Douglas J. Kennett and Bruce Winterhalder, eds., *Behavioral Ecology and the Transition to Agriculture* (Berkeley: University of California Press, 2006).

109 Lisa A. Maher, Tobias Richter, and Jay T. Stock, "The Pre-Natufian Epipaleolithic: Long-Term Behavioral Trends in the Levant," *Evolutionary Anthropology: Issues, News, and Reviews*, 21 (2012), 69–81; Anna Belfer-Cohen, "The Natufian in the Levant," *Annual Review of Anthropology*, 20 (1991), 167–86.

110 Alan H. Simmons, *The Neolithic Revolution in the Near East: Transforming the Human Landscape* (Tucson: University of Arizona Press, 2011).

111 Massimo Livi Bacci, *A Concise History of World Population, 6th Edition* (West Sussex: John Wiley & Sons, 2017).

112 Lewis Mumford, *The City in History: Its Origins, Its Transformations, and Its Prospects* (New York: Houghton Mifflin Harcourt, 1961).

113 Azar Gat, *War in Human Civilization* (Oxford: Oxford University Press, 2008).

114 Glenn E. Perry, *The History of Egypt* (Santa Barbara: ABC-CLIO, 2015).

115 Peter R. Bedford, "The Neo-Assyrian Empire," in *The Dynamics of Ancient Empires: State Power from Assyria to Byzantium*, ed. Ian Morris and Walter Scheidel (Oxford: Oxford University Press, 2009), 30–65; Tony J. Wilkinson, Jason Ur, Eleanor Barbanes Wilkinson and Mark Altaweel, "Landscape and Settlement in the Neo-Assyrian Empire," *Bulletin of the American Schools of Oriental Research*, 340 (2005), 23–56; Peter Sarris, "Introduction: Aristocrats, Peasants and the Transformation of Rural Society, c. 400–800," *Journal of Agrarian Change*, 9 (2009), 3–22.

116 John H. Kautsky, *The Politics of Aristocratic Empires* (London and New York: Routledge, 2017).

117 Wayne M. Senner, *The Origins of Writing* (Lincoln and London: University of Nebraska Press, 1991).

118 Rosalind Thomas, *Oral Tradition and Written Record in Classical Athens* (Cambridge: Cambridge University Press, 1989).

119 William G. Boltz, "The Invention of Writing in China," *Oriens Extremus*, 42 (2000), 1–17.

120 John D. Ray, "The Emergence of Writing in Egypt," *World Archaeology*, 17 (1986), 307–16.

121 Shawki M. Farag, "Ancient Egypt: The Development of Record Keeping in the 'Old Kingdom,'" *Accounting & Financial History Research Journal* (2012).

122 Valentine Roux and M.-A. Courty, "Identification of Wheel-Fashioning Methods: Technological Analysis of 4th–3rd Millennium BC Oriental Ceramics," *Journal of Archaeological Science*, 25 (1998), 747–63.

123 Glyn Davies, *History of Money* (Cardiff: University of Wales Press, 2010).

124 Pamela Nightingale, "Monetary Contraction and Mercantile Credit in Later Medieval England 1," *The Economic History Review*, 43 (1990), 560–75.

125 Robert Henry Pfeiffer, "An Analysis of the Hammurabi Code," *The American Journal of Semitic Languages and Literatures*, 36 (1920), 310–15; K. V. Nagarajan, "The Code of Hammurabi: An Economic Interpretation," *International Journal of Business and Social Science*, 2 (2011), 108–16.

126 Alfred Kieser, "Organizational, Institutional, and Societal Evolution: Medieval Craft Guilds and the Genesis of Formal Organizations," *Administrative Science Quarterly* (1989), 540–64; Emily Kadens, "Order within Law, Variety within Custom: The Character of the Medieval Merchant Law," *Chicago Journal of International Law*, 5 (2004), 39–65.

127 F. Rachel Magdalene and Cornelia Wunsch, "Slavery between Judah and Babylon: The Exilic Experience," in *Slaves and Households in the Near East*, ed. L. Culbertson (Chicago: OIS, 2011), 113–34.

128 Kostas Vlassopoulos, "Greek Slavery: From Domination to Property and Back Again," *The Journal of Hellenic Studies*, 131 (2011), 115–30.

129 David Herlihy, *The History of Feudalism* (New York: Springer, 1971).

130 Rosamond Jane Faith, "Peasant Families and Inheritance Customs in Medieval England," *The Agricultural History Review*, 14 (1966), 77–95.

131 Joshua C. Tate, "Inheritance Rights of Nonmarital Children in Late Roman Law," *Roman Legal Tradition*, 4 (2008), 1–36.

132 Laura Betzig, "Medieval Monogamy," *Journal of Family History*, 20 (1995), 181–216.

133 David Schaps, "Women in Greek Inheritance Law," *The Classical Quarterly*, 25 (1975), 53–57.

134 Tikva Simone Frymer-Kensky, *In the Wake of the Goddesses: Women, Culture, and the Biblical Transformation of Pagan Myth* (New York and Toronto: Free Press, 1992).

135 Sabine R. Huebner, "Egypt as Part of the Mediterranean? Domestic Space and Household Structures in Roman Egypt," in *Mediterranean Families in Antiquity: Households, Extended Families, and Domestic Space*, ed. Sabine R. Huebner and Geoffrey Nathan (Malden, MA: Wiley-Blackwell, 2017), 154–73.

136 Sabine R. Huebner, "A Mediterranean Family? A Comparative Approach to the Ancient World," in *Mediterranean Families in Classical Antiquity: Households, Extended Families, and Domestic Space*, ed. Sabine R. Huebner and Geoffrey Nathan (Malden, MA: Wiley-Blackwell, 2017), 3–26.

137 Steven Ruggles, "Multigenerational Families in Nineteenth-Century America," *Continuity and Change*, 18 (2003), 139–65.

138 Michael Anderson, *Approaches to the History of the Western Family 1500–1914*, vol. 1 (Cambridge: Cambridge University Press, 1995).

139 S. Tedeschi, "Wordsworth and the Affects of Urbanization," in *Urbanization and English Romantic Poetry* (Cambridge: Cambridge University Press, 2017), 76–110.

Chapter 3

140 Patrick K. O'Brien, "Agriculture and the Industrial Revolution," *The Economic History Review*, 30 (1977), 166–81; Robert C. Allen, *The Industrial Revolution: A Very Short Introduction* (Oxford: Oxford University Press, 2017).

141 Jacob L. Weisdorf, "From Domestic Manufacture to Industrial Revolution: Long-Run Growth and Agricultural Development," *Oxford Economic Papers*, 58 (2006), 264–87.

142 Andrew Lees and Lynn Hollen Lees, *Cities and the Making of Modern Europe, 1750–1914* (Cambridge: Cambridge University Press, 2007).

143 Danielle Thorne, *Hidden in History: The Untold Stories of Women during the Industrial Revolution* (Ocala, FL: Atlantic Publishing Company, 2019); Anand A. Yang, "Peasants on the Move: A Study of Internal Migration in India," *The Journal of Interdisciplinary History*, 10 (1979), 37–58.

144 Peter F. Drucker, *The New Society: The Anatomy of Industrial Order* (New York: Routledge, 2017).

145 Jack Goody, *The Development of the Family and Marriage in Europe* (Cambridge: Cambridge University Press, 1983); Steven Mintz and Susan Kellogg, *Domestic Revolutions: A Social History of American Family Life* (New York: Simon and Schuster, 1989); Stephen Pridgen and Donald Weinbrenner, "Industrialization and Family Change," in *Encyclopedia of Family Studies*, ed. C. L. Shehan (New York: John Wiley & Sons, 2016), 1–6.

146 Steven Ruggles, "The Decline of Intergenerational Coresidence in the United States, 1850 to 2000," *American Sociological Review*, 72 (2007), 964–89.

147 Steven Ruggles, "Reconsidering the Northwest European Family System: Living Arrangements of the Aged in Comparative Historical Perspective," *Population and Development Review*, 35 (2009), 249–73.

148 Joseph Agassi, *The Very Idea of Modern Science: Francis Bacon and Robert Boyle* (New York: Springer, 2012).

149 Clark Glymour, Richard Scheines, and Peter Spirtes, *Discovering Causal Structure: Artificial Intelligence, Philosophy of Science, and Statistical Modeling* (Orlando, FL: Academic Press, 2014).

150 Richard Olson, *Science and Religion, 1450–1900: From Copernicus to Darwin* (Westport: Greenwood Publishing Group, 2004).

151 Thomas Deissinger, "The Evolution of the Modern Vocational Training Systems in England and Germany: A Comparative View," *Compare*, 24 (1994), 17–36; Bruno Belhoste and Konstantinos Chatzis, "From Technical Corps to Technocratic Power: French State Engineers and their Professional and Cultural Universe in the First Half of the 19th Century," *History and Technology*, 23 (2007), 209–25.

152 Nicholas Crafts, "The First Industrial Revolution: Resolving the Slow Growth/Rapid Industrialization Paradox," *Journal of the European Economic Association*, 3 (2005), 525–34; Naomi R. Lamoreaux and Kenneth L. Sokoloff, "Market Trade in Patents and the Rise of a Class of Specialized Inventors in the 19th-Century United States," *American Economic Review*, 91 (2001), 39–44.

153 Robert C. Allen, *Global Economic History: A Very Short Introduction*, vol. 282 (Oxford: Oxford University Press, 2011).

154 Andrew Atkeson and Patrick J. Kehoe, "The Transition to a New Economy after the Second Industrial Revolution" (National Bureau of Economic Research, 2001).

155 Eric L. Jones, "The Agricultural Labour Market in England, 1793–1872," *The Economic History Review*, 17 (1964), 322–38.

156 Wolfgang Schivelbusch, *Disenchanted Night: The Industrialization of Light in the Nineteenth Century* (Berkeley: University of California Press, 1995).

157 Michael Huberman, Christopher M. Meissner, and Kim Oosterlinck, "Technology and Geography in the Second Industrial Revolution: New Evidence from the Margins of Trade," *The Journal of Economic History*, 77 (2017), 39–89.

158 Michael F. Graham, *Blasphemies of Thomas Aikenhead: Boundaries of Belief on the Eve of the Enlightenment* (Edinburgh: Edinburgh University Press,

2008); Dorinda Outram, *The Enlightenment*, vol. 58 (Cambridge: Cambridge University Press, 2019).

159 Mara Squicciarini, "Devotion and Development: Religiosity, Education, and Economic Progress in 19th-Century France," *Education, and Economic Progress in 19th-Century France (July 2019). CEPR Discussion Paper No. DP13877* (2019).

160 Michael W. McConnell, "God Is Dead and We Have Killed Him: Freedom of Religion in the Post-Modern Age," *Brigham Young University Law Review* (1993), 163–88.

161 Paul Puschmann and Arne Solli, "Household and Family during Urbanization and Industrialization: Efforts to Shed New Light on an Old Debate," *The History of the Family*, 19 (2014), 1–12.

162 Rolf Gehrmann, "German Towns at the Eve of Industrialization: Household Formation and the Part of the Elderly," *The History of the Family*, 19 (2014), 13–28.

163 Edward Shorter, *The Making of the Modern Family* (New York: Basic Books, 1975).

164 Tea Moon Seok and Ho Chol Lee, "A Study of Fertilization System in the 18th Century," *Current Research on Agriculture and Life Sciences*, 5 (1987), 205–20; D. Gale Johnson, "Agriculture and the Wealth of Nations," *The American Economic Review*, 87 (1997), 1–12.

165 Liam Harte, "John O'neill, 'Fifty Years' Experience of an Irish Shoemaker in London,'" in *The Literature of the Irish in Britain* (New York: Springer, 2009), 15–19.

166 Mattei Dogan and John D. Kasarda, *The Metropolis Era: A World of Giant Cities*, vol. 1 (New York: Sage, 1988).

167 Alan Gilbert and Josef Gugler, *Cities Poverty and Development: Urbanization in the Third World*, 2nd ed. (Oxford: Oxford University, 1992).

168 Mattei Dogan, "Giant Cities as Maritime Gateways," in *The Metropolis Era: A World of Giant Cities*, ed. M. Dogan and J. D. Kasarda (New York: Sage, 1988), 30–55.

169 Konrad H. Jarausch and Wilhelm H. Schröder, *Quantitative History of Society and Economy: Some International Studies*, vol. 21 (St Katharinen: Scripta Mercaturae Verl., 1987).

170 Karin Kurz, *Home Ownership and Social Inequality in a Comparative Perspective* (Stanford: Stanford University Press, 2004).

171 Hartmut Kaelble, "Social Inequality in the 19th and 20th Centuries: Some Introductory Remarks," in *Quantitative History of Society and Economy: Some International Studies*, ed. K. H. Jarausch and W. H. Schröder (St Katharinen: Scripta Mercaturae Verl., 1987).

172 Frances Holliss, *Beyond Live/Work: The Architecture of Home-Based Work* (New York: Routledge, 2015).

173 Frances Trollope, *The Life and Adventures of Michael Armstrong: The Factory Boy* (London: Frank Cass and Company Limited, 1840).

174 National Bureau of Statistics of China, "China Statistics: National Statistics" (Beijing: 2013).

175 Montagu Stephen Williams, *Later Leaves: Being the Further Reminiscences of Montagu Williams, Qc* (London: Macmillan, 1893).

176 Dominic A. Pacyga, *Chicago: A Biography* (Chicago: University of Chicago Press, 2009).

177 Frank F. Furstenberg, "Family Change in Global Perspective: How and Why Family Systems Change," *Family Relations*, 68 (2019), 326–41.

178 Stephanie Coontz, *Marriage, a History: How Love Conquered Marriage* (New York: Penguin, 2006).

179 Arland Thornton, "The Developmental Paradigm, Reading History Sideways, and Family Change," *Demography*, 38 (2001), 449–65; Tamara K. Hareven, "The History of the Family and the Complexity of Social Change," *The American Historical Review*, 96 (1991), 95–124.

180 Linda G. Martin, "Changing Intergenerational Family Relations in East Asia," *The Annals of the American Academy of Political and Social Science*, 510 (1990), 102–14; John Knodel and Nibhon Debavalya, "Living Arrangements and Support among the Elderly in South-East Asia: An Introduction," *Asia Pacific Population Journal*, 12 (1997), 5–16; Albert I. Hermalin, Mary Beth Ofstedal, and Ming-Cheng Chang, "Types of Supports for the Aged and their Providers in Taiwan," *Aging and Generational Relations over the Life Course: A Historical and Cross-Cultural Perspective* (New York: Walter de Gruyter, 1996), 400–37; David I. Kertzer, Dennis P. Hogan, Robert W. Freckmann, and Lynn G. Clark, *Family, Political Economy, and Demographic Change: The Transformation of Life in Casalecchio, Italy, 1861–1921* (Madison: University of Wisconsin Press, 1989).

181 Jean Louis Flandrin, *Families in Former Times* (London and New York: Cambridge Press, 1979)

182 Frederic Le-Play, *Les Ouvriers Européens: L'organisation Des Familles [European Laborers: The Organization of Families]* (Paris: Imprimerie imperiale, 1855).

183 Emile Durkheim, "Introduction À La Sociologie De La Famille," in *Annales de la Faculté des Lettres de Bordeaux (257–281)*. Translated in Mark Traugott, *Émile Durkheim on Institutional Analysis*, 205–228 (Chicago: University of Chicago Press, 1888).

184 Ernest Watson Burgess, *The Function of Socialization in Social Evolution* (Chicago: University of Chicago Press, 1916); William F. Ogburn, *The Family and its Functions: Report of the President's Research Committee on Social Trends, Recent Social Trends in the United States* (New York: McGraw Hill, 1934); Talcott Parsons, "The Social Structure of the Family," in *The Family: Its Function and Destiny*, ed. R. N. Anshen (New York: Harper, 1949), 173–201.

185 Robert E. Park, "Human Migration and the Marginal Man," *American Journal of Sociology*, 33 (1928), 881–93; Robert E. Park, "The City: Suggestions for the Investigation of Human Behavior in the City Environment," *American Journal of Sociology*, 20 (1915), 577–612; William Isaac Thomas and Florian Znaniecki,

The Polish Peasant in Europe and America: Monograph of an Immigrant Group (Chicago: University of Chicago Press, 1918).

186 Jérôme Bourdieu and Lionel Kesztenbaum, " 'The True Social Molecule.' Industrialization, Paternalism and the Family. Half a Century in Le Creusot (1836–86)," *The History of the Family*, 19 (2014), 53–76.

Chapter 4

187 Diane Singerman, "Youth, Gender, and Dignity in the Egyptian Uprising," *Journal of Middle East Women's Studies*, 9 (2013), 1–27; Zeynep Tufekci and Christopher Wilson, "Social Media and the Decision to Participate in Political Protest: Observations from Tahrir Square," *Journal of Communication*, 62 (2012), 363–79.

188 Richard G. Harris, "The Knowledge-Based Economy: Intellectual Origins and New Economic Perspectives," *International Journal of Management Reviews*, 3 (2001), 21–40; Lester C. Thurow, "Globalization: The Product of a Knowledge-Based Economy," *The Annals of the American Academy of Political and Social Science*, 570 (2000), 19–31; Loet Leydesdorff, *The Knowledge-Based Economy: Modeled, Measured, Simulated* (Boca Raton: Universal-Publishers, 2006).

189 Hunt Allcott and Matthew Gentzkow, "Social Media and Fake News in the 2016 Election," *Journal of Economic Perspectives*, 31 (2017), 211–36; David M. J. Lazer, Matthew A. Baum, Yochai Benkler, Adam J. Berinsky, Kelly M. Greenhill, Filippo Menczer, Miriam J. Metzger, Brendan Nyhan, Gordon Pennycook, and David Rothschild, "The Science of Fake News," *Science*, 359 (2018), 1094–96.

190 R. D. Putnam, *Bowling Alone: The Collapse and Revival of American Community* (New York: Simon and Schuster, 2000); Lee Raine and Barry Wellman, "Networked," *The New Social Operating System* (Cambridge, MA: MIT Press, 2012).

191 Manuel Castells, *The Rise of the Network Society*, 2nd ed. (Malden, MA: Wiley-Blackwell, 2010).

192 Elyakim Kislev, *Happy Singlehood: The Rising Acceptance and Celebration of Solo Living* (Oakland: University of California Press, 2019).

193 Stephanie Coontz, *The Way We Never Were: American Families and the Nostalgia Trap* (New York: Basic Books, 2016).

194 Eric Klinenberg, *Going Solo: The Extraordinary Rise and Surprising Appeal of Living Alone* (New York: Penguin, 2012).

195 Elyakim Kislev, "Happiness, Post-Materialist Values, and the Unmarried," *Journal of Happiness Studies*, 19 (2018), 2243–65.

196 Joseph Migga Kizza, *Ethical and Social Issues in the Information Age*, 6th ed. (New York: Springer, 2017).

197 Paul DiMaggio, Eszter Hargittai, W. Russell Neuman, and John P. Robinson, "Social Implications of the Internet," *Annual Review of Sociology*, 27 (2001), 307–36.

198 Luciano Floridi, *Information: A Very Short Introduction* (Oxford: Oxford University Press, 2010).

199 Don Mennie, "Computers: Personal Computers for the Entrepreneur: Microcomputers Can Help Meet Payrolls Rather Than Play Star Wars Availability of Applications Software Packages Is the Pacing Item," *IEEE Spectrum*, 15 (1978), 30–35.

200 Robert D. Atkinson and Andrew S. McKay, "Digital Prosperity: Understanding the Economic Benefits of the Information Technology Revolution," Available at SSRN 1004516 (2007).

201 Ed H. Chi, "The Social Web: Research and Opportunities," *Computer*, 41 (2008), 88–91.

202 Janet Abbate, *Inventing the Internet* (Cambridge, MA: MIT Press, 2000).

203 Jeff Desjardins, "This Is What Happens in a Minute on the Internet" (Visual Capitalist & World Economic Forum, 2020).

204 Desmond Lobo, Kerem Kaskaloglu, Cha Young Kim, and Sandra Herbert, "Web Usability Guidelines for Smartphones: A Synergic Approach," *International Journal of Information and Electronics Engineering*, 1 (2011), 33–37.

205 Martin Campbell-Kelly and Daniel D. Garcia-Swartz, *From Mainframes to Smartphones: A History of the International Computer Industry* (Cambridge, MA: Harvard University Press, 2015).

206 Robin Mansell, "The Life and Times of the Information Society," *Prometheus*, 28 (2010), 165–86.

207 Abraham Charnes, William W. Cooper, and Edwardo Rhodes, "Evaluating Program and Managerial Efficiency: An Application of Data Envelopment Analysis to Program Follow Through," *Management Science*, 27 (1981), 668–97.

208 Daniel Esty and Reece Rushing, "Governing by the Numbers: The Promise of Data-Driven Policymaking in the Information Age" (Washington, DC: Center for American Progress, 2007), 21; Patrick Dunleavy and Christopher Hood, "From Old Public Administration to New Public Management," *Public Money & Management*, 14 (1994), 9–16; Richard Woodward, "The Organisation for Economic Cooperation and Development: Global Monitor," *New Political Economy*, 9 (2004), 113–27.

209 Gartner, "Gartner Says Worldwide Data Center Infrastructure Spending to Grow 6% in 2021" (Stamford, CT: Gartner, 2020).

210 Yoneji Masuda, *The Information Society as Post-Industrial Society* (Washington, DC: World Future Society, 1981).

211 Anja Bechmann, "Internet Profiling: The Economy of Data Intraoperability on Facebook and Google," *MedieKultur: Journal of Media and Communication Research*, 29 (2013), 19.

212   Olivia J. Walch, Amy Cochran, and Daniel B. Forger, "A Global Quantification of "Normal" Sleep Schedules Using Smartphone Data," *Science Advances*, 2 (2016), e1501705; J. Clement, "Worldwide Digital Population as of October 2020" (Statista, 2020).

213   Yannis Theocharis, "The Conceptualization of Digitally Networked Participation," *Social Media + Society*, 1 (2015), 2056305115610140.

214   Sunil Setti and Anjar Wanto, "Analysis of Backpropagation Algorithm in Predicting the Most Number of Internet Users in the World," *Jurnal Online Informatika*, 3 (2019), 110–15.

215   Wenhong Chen and Barry Wellman, "Minding the Cyber-Gap: The Internet and Social Inequality," in *The Blackwell Companion to Social Inequalities*, ed. Mary Romero and Eric Margolis (Malden, MA: Blackwell Publishing, 2005), 523–45.

216   Jan Van Dijk, *The Network Society* (Thousand Oaks, CA: Sage Publications, 2012).

217   Candice L. Odgers and Michaeline R. Jensen, "Annual Research Review: Adolescent Mental Health in the Digital Age: Facts, Fears, and Future Directions," *Journal of Child Psychology and Psychiatry*, 61 (2020), 336–48.

218   V. Rideout and M. B. Robb, "Social Media, Social Life: Teens Reveal their Experiences" (San Francisco, CA: Common Sense Media, 2018).

219   Larry D. Rosen, *Rewired: Understanding the Igeneration and the Way They Learn* (New York: St. Martin's Press, 2010); Jean M. Twenge, *Igen: Why Today's Super-Connected Kids Are Growing Up Less Rebellious, More Tolerant, Less Happy—and Completely Unprepared for Adulthood—and What That Means for the Rest of Us* (New York: Simon and Schuster, 2017).

220   Jing Zhang and Sharon Shavitt, "Cultural Values in Advertisements to the Chinese X-Generation—Promoting Modernity and Individualism," *Journal of Advertising*, 32 (2003), 23–33; Henri C. Santos, Michael E. W. Varnum, and Igor Grossmann, "Global Increases in Individualism," *Psychological Science*, 28 (2017), 1228–39.

221   Sara H. Konrath, Edward H. O'Brien, and Courtney Hsing, "Changes in Dispositional Empathy in American College Students over Time: A Meta-Analysis," *Personality and Social Psychology Review*, 15 (2011), 180–98.

222   Manolo Farci, Luca Rossi, Giovanni Boccia Artieri, and Fabio Giglietto, "Networked Intimacy. Intimacy and Friendship among Italian Facebook Users," *Information, Communication & Society*, 20 (2017), 784–801; Valerie Francisco, " 'The Internet Is Magic': Technology, Intimacy and Transnational Families," *Critical Sociology*, 41 (2015), 173–90.

223   Mitchell Hobbs, Stephen Owen, and Livia Gerber, "Liquid Love? Dating Apps, Sex, Relationships and the Digital Transformation of Intimacy," *Journal of Sociology*, 53 (2017), 271–84.

224   Matteo Marsili, Fernando Vega-Redondo, and František Slanina, "The Rise and Fall of a Networked Society: A Formal Model," *Proceedings of the National Academy of Sciences*, 101 (2004), 1439–42.

225 Barry Wellman, Anabel Quan-Haase, Jeffrey Boase, Wenhong Chen, Keith Hampton, Isabel Díaz, and Kakuko Miyata, "The Social Affordances of the Internet for Networked Individualism," *Journal of Computer-Mediated Communication*, 8 (2003), JCMC834.

226 Chung-tai Cheng, "The Flow of Information as Empowerment and the Changing Social Networking Landscape in Rural China," *Library Trends*, 62 (2013), 81–94.

227 Barry Wellman, "Computer Networks as Social Networks," *Science*, 293 (2001), 2031–34.

228 Colin Campbell, *The Romantic Ethic and the Spirit of Modern Consumerism* (London: WritersPrintShop, 2005); Sharon Boden, *Consumerism, Romance and the Wedding Experience* (London and New York: Palgrave Macmillan, 2003).

229 Wendy Wang and Kim C. Parker, *Record Share of Americans Have Never Married: As Values, Economics and Gender Patterns Change, Social & Demographic Trends Project* (Washington, DC: Pew Research Center, 2014).

230 Euromonitor, *Downsizing Globally: The Impact of Changing Household Structure on Global Consumer Markets* (2013).

231 Premchand Dommaraju, "One-Person Households in India," *Demographic Research*, 32 (2015); Hyunjoon Park and Jaesung Choi, "Long-Term Trends in Living Alone among Korean Adults: Age, Gender, and Educational Differences," *Demographic Research*, S15 (2015), 1177–208; Christophe Guilmoto and Myriam de Loenzien, "Emerging, Transitory or Residual? One-Person Households in Viet Nam," *Demographic Research*, S15 (2015), 1147–76; Chai Podhisita and Peter Xenos, "Living Alone in South and Southeast Asia: An Analysis of Census Data," *Demographic Research*, S15 (2015), 1113–46.

232 Martin Wolfe, "The Concept of Economic Sectors," *The Quarterly Journal of Economics*, 69 (1955), 402–20; Colin Clark, "The Conditions of Economic Progress," *The Conditions of Economic Progress* (London: Macmillan, 1967); Allen G. B. Fisher, *Clash of Progress and Security* (London: Macmillan, 1935).

233 Silvia Gorenstein and Ricardo Ortiz, "Natural Resources and Primary Sector-Dependent Territories in Latin America," *Area Development and Policy*, 3 (2018), 42–59.

234 Andrew Kilpatrick and Tony Lawson, "On the Nature of Industrial Decline in the UK," *Cambridge Journal of Economics*, 4 (1980), 85–102.

235 Makada Henry-Nickie, Kwadwo Frimpong, and Hao Sun, "Trends in the Information Technology Sector" (Washington, DC: Brookings Institute, 2019).

236 Zoltan Kenessey, "The Primary, Secondary, Tertiary and Quaternary Sectors of the Economy," *Review of Income and Wealth*, 33 (1987), 359–85.

237 Nuša Erman, Katarina Rojko, and Dušan Lesjak, "Traditional and New CT Spending and its Impact on Economy," *Journal of Computer Information Systems* (2020), 1–13.

238 Kevin Barefoot, Dave Curtis, William Jolliff, Jessica R. Nicholson, and Robert Omohundro, "Defining and Measuring the Digital Economy," *US Department of Commerce Bureau of Economic Analysis, Washington, DC*, 15

(2018); Oxford Economics, "Digital Spillover: Measuring the True Impact of the Digital Economy," A Report by Huawei and Oxford Economics, Oxford, United Kingdom, https://www. oxfordeconomics.com/recentreleases/digital-spillover (2017).

239 Jong-Wha Lee and Hanol Lee, "Human Capital in the Long Run," *Journal of Development Economics*, 122 (2016), 147–69.

240 Yoav Lavee and Ruth Katz, "The Family in Israel: Between Tradition and Modernity," *Marriage & Family Review*, 35 (2003), 193–217.

241 Eli Berman, "Sect, Subsidy, and Sacrifice: An Economist's View of Ultra-Orthodox Jews" (Jerusalem: National Bureau of Economic Research, 1998); Tally Katz-Gerro, Sharon Raz, and Meir Yaish, "How Do Class, Status, Ethnicity, and Religiosity Shape Cultural Omnivorousness in Israel?," *Journal of Cultural Economics*, 33 (2009), 1–17.

242 Joelle Abramowitz, "Turning Back the Ticking Clock: The Effect of Increased Affordability of Assisted Reproductive Technology on Women's Marriage Timing," *Journal of Population Economics*, 27 (2014), 603–33; Ya'arit Bokek-Cohen and Limor Dina Gonen, "Sperm and Simulacra: Emotional Capitalism and Sperm Donation Industry," *New Genetics and Society*, 34 (2015), 243–73.

243 Elyakim Kislev, "How Do Relationship Desire and Sociability Relate to Each Other among Singles? Longitudinal Analysis of the Pairfam Survey," *Journal of Social and Personal Relationships* (2020); Elyakim Kislev, "Social Capital, Happiness, and the Unmarried: A Multilevel Analysis of 32 European Countries," *Applied Research in Quality of Life* 15 (2019), 1475–92.

244 Paul R. Amato, "Research on Divorce: Continuing Trends and New Developments," *Journal of Marriage and Family*, 72 (2010), 650–66; Cezar Santos and David Weiss, "'Why Not Settle Down Already?': A Quantitative Analysis of the Delay in Marriage," *International Economic Review*, 57 (2016), 425–52.

Chapter 5

245 David McCullough, *The Wright Brothers* (New York: Simon and Schuster, 2015).

246 Tuka Al Hanai, Mohammad M. Ghassemi, and James R. Glass, "Detecting Depression with Audio/Text Sequence Modeling of Interviews," in *Interspeech* (2018), 1716–20.

247 John Fox, "Towards a Canonical Theory of General Intelligence," *Journal of Artificial General Intelligence*, 11 (2020), 35–40.

248 Katja Grace, John Salvatier, Allan Dafoe, Baobao Zhang, and Owain Evans, "When Will AI Exceed Human Performance? Evidence from AI Experts," *Journal of Artificial Intelligence Research*, 62 (2018), 729–54.

249 Reid McIlroy-Young, Siddhartha Sen, Jon Kleinberg, and Ashton Anderson, "Aligning Superhuman AI with Human Behavior: Chess as a Model System," in

Proceedings of the 26th ACM SIGKDD International Conference on Knowledge Discovery & Data Mining (2020), 1677–87.

250   Feifei Shi, Huansheng Ning, Wei Huangfu, Fan Zhang, Dawei Wei, Tao Hong, and Mahmoud Daneshmand, "Recent Progress on the Convergence of the Internet of Things and Artificial Intelligence," IEEE Network, 34 (2020), 8–15.

251   Grace A. Martin, "For the Love of Robots: Posthumanism in Latin American Science Fiction between 1960–1999" (PhD diss., University of Kentucky, 2015).

252   Ayala Malakh-Pines, Falling in Love: Why We Choose the Lovers We Choose (New York: Taylor & Francis, 2005).

253   Gillian Stevens, Dawn Owens, and Eric C. Schaefer, "Education and Attractiveness in Marriage Choices," Social Psychology Quarterly (1990), 62–70; Matthijs Kalmijn, "Intermarriage and Homogamy: Causes, Patterns, Trends," Annual Review of Sociology, 24 (1998), 395–421.

254   Edelman, "2019 Edelamn AI Survey Results Report" (Berlin: Edelman AI Institute, 2019).

255   William James, The Varieties of Religious Experience (Cambridge, MA: Harvard University Press, 1985).

Chapter 6

256   Mike Murphy, "This App Is Trying to Replicate You" (2019), https://qz.com/1698337/replika-this-app-is-trying-to-replicate-you/.

257   Amir Ziv and Diana Bahur-Nir, "'Understanding Language Is the Next Big Challenge for AI,' Says Amnon Shashua" (Calcalist Ctech, 2020).

258   Korok Sengupta, Sayan Sarcar, Alisha Pradhan, Roisin McNaney, Sergio Sayago, Debaleena Chattopadhyay, and Anirudha Joshi, "Challenges and Opportunities of Leveraging Intelligent Conversational Assistant to Improve the Well-Being of Older Adults," in Extended Abstracts of the 2020 CHI Conference on Human Factors in Computing Systems (2020), 1–4; Jarosław Kowalski, Anna Jaskulska, Kinga Skorupska, Katarzyna Abramczuk, Cezary Biele, Wiesław Kopeć, and Krzysztof Marasek, "Older Adults and Voice Interaction: A Pilot Study with Google Home," in Extended Abstracts of the 2019 CHI Conference on Human Factors in Computing Systems (2019), 1–6.

259   Heetae Yang and Hwansoo Lee, "Understanding User Behavior of Virtual Personal Assistant Devices," Information Systems and e-Business Management, 17 (2019), 65–87; Amanda Purington, Jessie G. Taft, Shruti Sannon, Natalya N. Bazarova, and Samuel Hardman Taylor, "'Alexa Is My New Bff' Social Roles, User Satisfaction, and Personification of the Amazon Echo," in Proceedings of the 2017 CHI Conference Extended Abstracts on Human Factors in Computing Systems (2017), 2853–59.

260   John Haugeland, Artificial Intelligence: The Very Idea (Cambridge, MA: MIT Press, 1989).

261 Yann LeCun, Yoshua Bengio, and Geoffrey Hinton, "Deep Learning," *Nature*, 521 (2015), 436–44; Mark F. St John, and James L. McClelland, "Learning and Applying Contextual Constraints in Sentence Comprehension," *Artificial Intelligence*, 46 (1990), 217–57.

262 Yann LeCun, Bernhard Boser, John S. Denker, Donnie Henderson, Richard E. Howard, Wayne Hubbard, and Lawrence D. Jackel, "Backpropagation Applied to Handwritten Zip Code Recognition," *Neural Computation*, 1 (1989), 541–51; Thomas Serre, Lior Wolf, Stanley Bileschi, Maximilian Riesenhuber, and Tomaso Poggio, "Robust Object Recognition with Cortex-Like Mechanisms," *IEEE Transactions on Pattern Analysis and Machine Intelligence*, 29 (2007), 411–26.

263 Richard S. Sutton and Andrew G. Barto, *Reinforcement Learning: An Introduction* (Cambridge, MA: MIT Press, 2018).

264 Volodymyr Mnih, Koray Kavukcuoglu, David Silver, Andrei A. Rusu, Joel Veness, Marc G. Bellemare, Alex Graves, Martin Riedmiller, Andreas K. Fidjeland, and Georg Ostrovski, "Human-Level Control through Deep Reinforcement Learning," *Nature*, 518 (2015), 529–33.

265 Brenden M. Lake, Ruslan Salakhutdinov, and Joshua B. Tenenbaum, "Human-Level Concept Learning through Probabilistic Program Induction," *Science*, 350 (2015), 1332–38; Aaron van den Oord, Sander Dieleman, Heiga Zen, Karen Simonyan, Oriol Vinyals, Alex Graves, Nal Kalchbrenner, Andrew Senior, and Koray Kavukcuoglu, "Wavenet: A Generative Model for Raw Audio," *arXiv preprint arXiv:1609.03499* (2016).

266 MyHeritage; https://www.myheritage.co.il/.

267 Jianfeng Gao, Michel Galley, and Lihong Li, *Neural Approaches to Conversational AI: Question Answering, Task-Oriented Dialogues and Social Chatbots* (Boston: Now Foundations and Trends, 2019).

268 Tirin Moore and Marc Zirnsak, "Neural Mechanisms of Selective Visual Attention," *Annual Review of Psychology*, 68 (2017), 47–72.

269 Emilio Salinas and L. F. Abbott, "Invariant Visual Responses from Attentional Gain Fields," *Journal of Neurophysiology*, 77 (1997), 3267–72.

270 Jimmy Ba, Volodymyr Mnih, and Koray Kavukcuoglu, "Multiple Object Recognition with Visual Attention," *arXiv preprint arXiv:1412.7755* (2014); Kelvin Xu, Jimmy Ba, Ryan Kiros, Kyunghyun Cho, Aaron Courville, Ruslan Salakhudinov, Rich Zemel, and Yoshua Bengio, "Show, Attend and Tell: Neural Image Caption Generation with Visual Attention," in *International Conference on Machine Learning* (2015), 2048–57.

271 Dzmitry Bahdanau, Kyunghyun Cho, and Yoshua Bengio, "Neural Machine Translation by Jointly Learning to Align and Translate," *arXiv preprint arXiv:1409.0473* (2014); Alex Graves, Greg Wayne, Malcolm Reynolds, Tim Harley, Ivo Danihelka, Agnieszka Grabska-Barwińska, Sergio Gómez Colmenarejo, Edward Grefenstette, Tiago Ramalho, and John Agapiou, "Hybrid Computing Using a Neural Network with Dynamic External Memory," *Nature*, 538 (2016), 471–76.

272   Endel Tulving, "How Many Memory Systems Are There?," *American Psychologist*, 40 (1985), 385.

273   Charles R. Gallistel and Adam Philip King, *Memory and the Computational Brain: Why Cognitive Science Will Transform Neuroscience* (New York: John Wiley & Sons, 2011); Endel Tulving, "Episodic Memory: From Mind to Brain," *Annual Review of Psychology*, 53 (2002), 1–25; Larry R. Squire, Craig E. L. Stark, and Robert E. Clark, "The Medial Temporal Lobe," *Annu. Rev. Neurosci.*, 27 (2004), 279–306.

274   Charles Blundell, Benigno Uria, Alexander Pritzel, Yazhe Li, Avraham Ruderman, Joel Z. Leibo, Jack Rae, Daan Wierstra, and Demis Hassabis, "Model-Free Episodic Control," *arXiv preprint arXiv:1606.04460* (2016).

275   Máté Lengyel and Peter Dayan, "Hippocampal Contributions to Control: The Third Way," in *Advances in Neural Information Processing Systems* (2008), 889–96.

276   Demis Hassabis and Eleanor A. Maguire, "Deconstructing Episodic Memory with Construction," *Trends in Cognitive Sciences*, 11 (2007), 299–306.

277   Mevlana Gemici, Chia-Chun Hung, Adam Santoro, Greg Wayne, Shakir Mohamed, Danilo J. Rezende, David Amos, and Timothy Lillicrap, "Generative Temporal Models with Memory," *arXiv preprint arXiv:1702.04649* (2017); Junhyuk Oh, Xiaoxiao Guo, Honglak Lee, Richard L. Lewis, and Satinder Singh, "Action-Conditional Video Prediction Using Deep Networks in Atari Games," in *Advances in Neural Information Processing Systems* (2015), 2863–71.

278   Matej Moravčík, Martin Schmid, Neil Burch, Viliam Lisý, Dustin Morrill, Nolan Bard, Trevor Davis, Kevin Waugh, Michael Johanson, and Michael Bowling, "Deepstack: Expert-Level Artificial Intelligence in Heads-Up No-Limit Poker," *Science*, 356 (2017), 508–13; David Silver, Aja Huang, Chris J. Maddison, Arthur Guez, Laurent Sifre, George Van Den Driessche, Julian Schrittwieser, Ioannis Antonoglou, Veda Panneershelvam, and Marc Lanctot, "Mastering the Game of Go with Deep Neural Networks and Tree Search," *Nature*, 529 (2016), 484.

279   Leon A. Gatys, Alexander S. Ecker, and Matthias Bethge, "A Neural Algorithm of Artistic Style," *arXiv preprint arXiv:1508.06576* (2015).

280   Zixiaofan Yang, Bingyan Hu, and Julia Hirschberg, "Predicting Humor by Learning from Time-Aligned Comments," in *Interspeech 2019* (Graz, Austria: Graz University of Technology, 2019), 496–500.

281   Pega, "What Consumers Really Think About AI: A Global Study" (Cambridge, MA: Pega, 2019).

282   Peter Gentsch, "Conversational AI: How (Chat) Bots Will Reshape the Digital Experience," in *AI in Marketing, Sales and Service* (New York: Springer, 2019), 81–125; James H. Moor, ed., *The Turing Test: The Elusive Standard of Artificial Intelligence*, ed. James H. Fetzer. Vol. 30: *Studies in Cognitive Systems* (Dordrecht: Kluwer Academic Publishers, 2003); Kevin Warwick and Huma Shah, "Can Machines Think? A Report on Turing Test Experiments at the

Royal Society," *Journal of Experimental & Theoretical Artificial Intelligence*, 28 (2016), 989–1007.

283   H. Jafarian, A. H. Taghavi, A. Javaheri, and R. Rawassizadeh, "Exploiting Bert to Improve Aspect-Based Sentiment Analysis Performance on Persian Language," *arXiv preprint arXiv:2012.07510* (2020); OpenAI, "Openai Api" (2020); Tom B. Brown, Benjamin Mann, Nick Ryder, Melanie Subbiah, Jared Kaplan, Prafulla Dhariwal, Arvind Neelakantan, Pranav Shyam, Girish Sastry, and Amanda Askell, "Language Models Are Few-Shot Learners," *arXiv preprint arXiv:2005.14165* (2020).

284   Noam Slonim, Yonatan Bilu, Carlos Alzate, Roy Bar-Haim, Ben Bogin, Francesca Bonin, Leshem Choshen, Edo Cohen-Karlik, Lena Dankin, Lilach Edelstein, Liat Ein-Dor, Roni Friedman-Melamed, Assaf Gavron, Ariel Gera, Martin Gleize, Shai Gretz, Dan Gutfreund, Alon Halfon, Daniel Hershcovich, Ron Hoory, Yufang Hou, Shay Hummel, Michal Jacovi, Charles Jochim, Yoav Kantor, Yoav Katz, David Konopnicki, Zvi Kons, Lili Kotlerman, Dalia Krieger, Dan Lahav, Tamar Lavee, Ran Levy, Naftali Liberman, Yosi Mass, Amir Menczel, Shachar Mirkin, Guy Moshkowich, Shila Ofek-Koifman, Matan Orbach, Ella Rabinovich, Ruty Rinott, Slava Shechtman, Dafna Sheinwald, Eyal Shnarch, Ilya Shnayderman, Aya Soffer, Artem Spector, Benjamin Sznajder, Assaf Toledo, Orith Toledo-Ronen, Elad Venezian, and Ranit Aharonov, "An Autonomous Debating System," *Nature*, 591 (2021), 379–84.

285   David Levy, 'Passing Turing's Intelligence Test: Rules and Results' (2020), http://dev.kamicomputing.com/blog/passing-turings-intelligent-test-rules-and-results.

286   Gabriella M. Harari, Sandrine R. Müller, Clemens Stachl, Rui Wang, Weichen Wang, Markus Bühner, Peter J. Rentfrow, Andrew T. Campbell, and Samuel D. Gosling, "Sensing Sociability: Individual Differences in Young Adults' Conversation, Calling, Texting, and App Use Behaviors in Daily Life," *Journal of Personality and Social Psychology*, 119 (2020), 204–208.

287   Stephen Roller, Emily Dinan, Naman Goyal, Da Ju, Mary Williamson, Yinhan Liu, Jing Xu, Myle Ott, Kurt Shuster, and Eric M. Smith, "Recipes for Building an Open-Domain Chatbot," *arXiv preprint arXiv:2004.13637* (2020).

288   Karen Hao, "AI Still Gets Confused about How the World Works," *MIT Technology Review*, 123 (2020), 24–29.

289   Hojjat Abdollahi, Ali Mollahosseini, Josh T. Lane, and Mohammad H. Mahoor, "A Pilot Study on Using an Intelligent Life-Like Robot as a Companion for Elderly Individuals with Dementia and Depression," in *2017 IEEE-RAS 17th International Conference on Humanoid Robotics (Humanoids)* (IEEE, 2017), 541–46.

290   Aria Zand, Arjun Sharma, Zack Stokes, Courtney Reynolds, Alberto Montilla, Jenny Sauk, and Daniel Hommes, "An Exploration into the Use of a Chatbot for Patients with Inflammatory Bowel Diseases: Retrospective Cohort Study," *Journal of Medical Internet Research*, 22 (2020), e15589.

291 Adam Palanica, Peter Flaschner, Anirudh Thommandram, Michael Li, and Yan Fossat, "Physicians' Perceptions of Chatbots in Health Care: Cross-Sectional Web-Based Survey," *Journal of Medical Internet Research*, 21 (2019), e12887.

292 George A. Miller, "Wordnet: A Lexical Database for English," *Communications of the ACM*, 38 (1995), 39–41.

293 Abubakr Siddig and Andrew Hines, "A Psychologist Chatbot Developing Experience" (2019).

294 Carlos Toxtli, Andrés Monroy-Hernández, and Justin Cranshaw, "Understanding Chatbot-Mediated Task Management," in *Proceedings of the 2018 CHI Conference on Human Factors in Computing Systems* (2018), 1–6.

295 Vivian Ta, Caroline Griffith, Carolynn Boatfield, Xinyu Wang, Maria Civitello, Haley Bader, Esther DeCero, and Alexia Loggarakis, "User Experiences of Social Support from Companion Chatbots in Everyday Contexts: Thematic Analysis," *Journal of Medical Internet Research*, 22 (2020), e16235.

296 Joseph Weizenbaum, "Eliza—a Computer Program for the Study of Natural Language Communication between Man and Machine," *Communications of the ACM*, 9 (1966), 36–45.

297 David E. Hartman, "Artificial Intelligence or Artificial Psychologist? Conceptual Issues in Clinical Microcomputer Use," *Professional Psychology: Research and Practice*, 17 (1986), 528; David Servan-Schreiber, "Artificial Intelligence and Psychiatry," *Journal of Nervous and Mental Disease*, 174 (1986), 191–202.

298 David D. Luxton, "Artificial Intelligence in Psychological Practice: Current and Future Applications and Implications," *Professional Psychology: Research and Practice*, 45 (2014), 332.

299 Larry Zhang, Joshua Driscol, Xiaotong Chen, and Reza Hosseini Ghomi, "Evaluating Acoustic and Linguistic Features of Detecting Depression Sub-Challenge Dataset," in *Proceedings of the 9th International on Audio/Visual Emotion Challenge and Workshop* (2019), 47–53.

300 https://www.genderlessvoice.com.

301 Simon D'alfonso, Olga Santesteban-Echarri, Simon Rice, Greg Wadley, Reeva Lederman, Christopher Miles, John Gleeson, and Mario Alvarez-Jimenez, "Artificial Intelligence-Assisted Online Social Therapy for Youth Mental Health," *Frontiers in Psychology*, 8 (2017), 796.

302 Jonathan Lazar, Jinjuan Heidi Feng, and Harry Hochheiser, *Research Methods in Human-Computer Interaction* (Burlington: Morgan Kaufmann, 2017).

303 Flávio Luis de Mello and Roberto Lins de Carvalho, "Knowledge Geometry," *Journal of Information & Knowledge Management*, 14 (2015), 1550028.

304 Yi Mou and Kun Xu, "The Media Inequality: Comparing the Initial Human-Human and Human-AI Social Interactions," *Computers in Human Behavior*, 72 (2017), 432–40.

305 Leonhard Glomann, Viktoria Hager, Christian A. Lukas, and Matthias Berking, "Patient-Centered Design of an E-Mental Health App," in *International*

Conference on Applied Human Factors and Ergonomics (New York: Springer, 2018), 264–71.

306 Vesna Kirandziska and Nevena Ackovska, "A Concept for Building More Humanlike Social Robots and their Ethical Consequence," in *Proceedings of the International Conferences, ICT, Society and Human Beings (MCCSIS)* (2014), 15–19.

307 Nicu Sebe, Michael S. Lew, Yafei Sun, Ira Cohen, Theo Gevers, and Thomas S. Huang, "Authentic Facial Expression Analysis," *Image and Vision Computing*, 25 (2007), 1856–63.

308 Gerald Bieber, Niklas Antony, and Marian Haescher, "Touchless Heart Rate Recognition by Robots to Support Natural Human-Robot Communication," in *Proceedings of the 11th Pervasive Technologies Related to Assistive Environments Conference* (2018), 415–20; Marina Krakovsky, "News: Artificial (Emotional) Intelligence," *Communications of the ACM*, 61 (2018), 18–19.

309 Clifford I. Nass, Youngme Moon, and John Morkes, "Computers Are Social Actors: A Review of Current Research," in *Human Values and the Design of Computer Technology*, ed. Batya Friedman (Cambridge: Cambridge University Press, 1997), p. 137.

310 Minha Lee, Sander Ackermans, Nena van As, Hanwen Chang, Enzo Lucas, and Wijnand IJsselsteijn, "Caring for Vincent: A Chatbot for Self-Compassion," in *Proceedings of the 2019 CHI Conference on Human Factors in Computing Systems* (2019), 1–13.

311 Johan F. Hoorn, Elly A. Konijn, and Matthijs A. Pontier, "Dating a Synthetic Character Is Like Dating a Man," *International Journal of Social Robotics*, 11 (2019), 235–53.

312 https://interestingengineering.com/elon-musk-neuralink-cyberpunk-ai. Bergan, Brad. 2021. "Elon Musk Wants Neuralink Applicants to Make 'Cyberpunk Come True.'" *Interesting Engineering*.

313 Aaron Smith and Monica Anderson, *Automation in Everyday Life* (Washington, DC: Pew Research Center, Social & Demographic Trends Project, 2017).

Chapter 7

314 SuperData, "2019 Year in Review: Digital Games and Interactive Media" (New York: Nielsen, 2020).

315 Sam Byford, "The Oculus Quest 2 Was Preordered Five Times as Much as the Original," *The Verge* (October 30, 2020).

316 Chris Nuttall, "Amazon's Autonomous Drive," *Financial Times* (2020); Aodhan Gregory, "Xrspace Manova Review: A VR Headset That's 'the Closest We'll Get to a Fully Immersive Digital World,'" *The Mirror* (2021).

317 Jung So-yeon, "Hyperconnect to Launch the World's First 'Human-AI-Powered Social Network' This Year," *Korea IT Times* (2021), http://www.korea ittimes.com/news/articleView.html?idxno=102956.

318 Sabrina A. Huang, and Jeremy Bailenson, "Close Relationships and Virtual Reality," in *Mind, Brain and Technology* (Springer, 2019), 49–65.

319 Branden Thornhill-Miller, and Jean-Marc Dupont, "Virtual Reality and the Enhancement of Creativity and Innovation: Under Recognized Potential among Converging Technologies?," *Journal of Cognitive Education and Psychology*, 15 (2016), 102–21.

320 Samuel Greengard, *Virtual Reality* (Cambridge, MA: MIT Press, 2019).

321 Radwa Mabrook, and Jane B. Singer, "Virtual Reality, 360° Video, and Journalism Studies: Conceptual Approaches to Immersive Technologies," *Journalism Studies*, 20:14 (2019), 1–17.

322 Yurgos Politis, Louis Olivia, Thomas Olivia, and Connie Sung, "Involving People with Autism in Development of Virtual World for Provision of Skills Training," *International Journal of E-Learning & Distance Education*, 32 (2017), 1–16.

323 Donghee Shin, "Empathy and Embodied Experience in Virtual Environment: To What Extent Can Virtual Reality Stimulate Empathy and Embodied Experience?," *Computers in Human Behavior*, 78 (2018), 64–73.

324 William R. Sherman and Alan B. Craig, *Understanding Virtual Reality: Interface, Application, and Design* (Cambridge, MA: Morgan Kaufmann, 2018).

325 Morteza Dianatfar, Jyrki Latokartano, and Minna Lanz, "Review on Existing Vr/Ar Solutions in Human–Robot Collaboration," *Procedia CIRP*, 97 (2021), 407–11.

326 Tanya Basu, "Facebook Is Making a Bracelet That Lets You Control Computers with Your Brain," *MIT Technology Review*, 124 (2021).

327 Tina Iachini, Luigi Maffei, Massimiliano Masullo, Vincenzo Paolo Senese, Mariachiara Rapuano, Aniello Pascale, Francesco Sorrentino, and Gennaro Ruggiero, "The Experience of Virtual Reality: Are Individual Differences in Mental Imagery Associated with Sense of Presence?," *Cognitive Processing* 20 (2018), 1 8.

328 Sandra Maria Correia Loureiro, João Guerreiro, and Faizan Ali, "20 Years of Research on Virtual Reality and Augmented Reality in Tourism Context: A Text-Mining Approach," *Tourism Management*, 77 (2020), 1–21.

329 Pietro Cipresso, Irene Alice Chicchi Giglioli, Mariano Alcañiz Raya, and Giuseppe Riva, "The Past, Present, and Future of Virtual and Augmented Reality Research: A Network and Cluster Analysis of the Literature," *Frontiers in Psychology*, 9 (2018), 1–20.

330 Gary Bitter and Allen Corral, "The Pedagogical Potential of Augmented Reality Apps," *Journal of Engineering Science Invention*, 3 (2014), 13–17.

331 Johannes Meyer, Thomas Schlebusch, Wolfgang Fuhl, and Enkelejda Kasneci, "A Novel Camera-Free Eye Tracking Sensor for Augmented Reality Based on Laser Scanning," *IEEE Sensors Journal*, 20 (2020), 15204–12.

332 Ana R. C. Donati, Solaiman Shokur, Edgard Morya, Debora S. F. Campos, Renan C. Moioli, Claudia M. Gitti, Patricia B. Augusto, Sandra Tripodi, Cristhiane G. Pires, and Gislaine A. Pereira, "Long-Term Training with a Brain-Machine Interface-Based Gait Protocol Induces Partial Neurological Recovery in Paraplegic Patients," *Scientific Reports*, 6 (2016), 1–16.

333   Amol P. Yadav, Daniel Li, and Miguel A. L. Nicolelis, "A Brain to Spine Interface for Transferring Artificial Sensory Information," *Scientific Reports*, 10 (2020), 1–15.

334   Tianfang Yan, Seiji Kameda, Katsuyoshi Suzuki, Taro Kaiju, Masato Inoue, Takafumi Suzuki, and Masayuki Hirata, "Minimal Tissue Reaction after Chronic Subdural Electrode Implantation for Fully Implantable Brain–Machine Interfaces," *Sensors*, 21 (2021), 178.

335   Franziska Roesner, Tadayoshi Kohno, and David Molnar, "Security and Privacy for Augmented Reality Systems," *Communications of the ACM*, 57 (2014), 88–96.

336   Don Karl, Kirk Soderquist, Miriam Farhi, Andrew Grant, Brendan Murphy, Jason Schneiderman, and Ben Straughan, *2018 Augmented and Virtual Reality Survey Report* (Seattle: Perkins Coie, 2018).

337   Dean Takahashi, "The Deanbeat: It's Time to Hurry up and Build the Metaverse," *Venture Beat* (2020), https://venturebeat.com/2020/05/22/the-deanbeat-its-time-to-hurry-up-and-build-the-metaverse/.

338   Philip Lindner, "Better, Virtually: The Past, Present, and Future of Virtual Reality Cognitive Behavior Therapy," *International Journal of Cognitive Therapy*, 14 (2020), 1–24.

339   Charles Xueyang Lin, Chaiwoo Lee, Dennis Lally, and Joseph F. Coughlin, "Impact of Virtual Reality (VR) Experience on Older Adults' Well-Being," in *International Conference on Human Aspects of IT for the Aged Population*, ed. J. Zhou and G. Salvendy (New York: Springer, 2018), 89–100.

340   Javier Fernández-Álvarez, Daniele Di Lernia, and Giuseppe Riva, "Virtual Reality for Anxiety Disorders: Rethinking a Field in Expansion," in *Anxiety Disorders* (New York: Springer, 2020), 389–414.

341   Cristina Botella, Javier Fernández-Álvarez, Verónica Guillén, Azucena García-Palacios, and Rosa Baños, "Recent Progress in Virtual Reality Exposure Therapy for Phobias: A Systematic Review," *Current Psychiatry Reports*, 19 (2017), 42.

342   Albert "Skip" Rizzo and Russell Shilling, "Clinical Virtual Reality Tools to Advance the Prevention, Assessment, and Treatment of PTSD," *European Journal of Psychotraumatology*, 8 (2017), 1–20; Cristina Botella, Berenice Serrano, Rosa M. Baños, and Azucena Garcia-Palacios, "Virtual Reality Exposure-Based Therapy for the Treatment of Post-Traumatic Stress Disorder: A Review of its Efficacy, the Adequacy of the Treatment Protocol, and its Acceptability," *Neuropsychiatric Disease and Treatment*, 11 (2015), 2533.

343   JoAnn Difede, Judith Cukor, Katarzyna Wyka, Megan Olden, Hunter Hoffman, Francis S. Lee, and Margaret Altemus, "D-Cycloserine Augmentation of Exposure Therapy for Post-Traumatic Stress Disorder: A Pilot Randomized Clinical Trial," *Neuropsychopharmacology*, 39 (2014), 1052–58.

344   Caroline J. Falconer, Aitor Rovira, John A. King, Paul Gilbert, Angus Antley, Pasco Fearon, Neil Ralph, Mel Slater, and Chris R. Brewin, "Embodying Self-Compassion within Virtual Reality and Its Effects on Patients with Depression," *BJPsych open*, 2 (2016), 74–80.

345   Karen E. Gerdes, Elizabeth A. Segal, Kelly F. Jackson, and Jennifer L. Mullins, "Teaching Empathy: A Framework Rooted in Social Cognitive Neuroscience and Social Justice," *Journal of Social Work Education*, 47 (2011), 109–31; Alan K. Louie, John H. Coverdale, Richard Balon, Eugene V. Beresin, Adam M. Brenner, Anthony P. S. Guerrero, and Laura Weiss Roberts, "Enhancing Empathy: A Role for Virtual Reality?," *Academic Psychiatry*, 42 (2018), 747–52.

346   Emily Carl, Aliza T. Stein, Andrew Levihn-Coon, Jamie R. Pogue, Barbara Rothbaum, Paul Emmelkamp, Gordon J. G. Asmundson, Per Carlbring, and Mark B. Powers, "Virtual Reality Exposure Therapy for Anxiety and Related Disorders: A Meta-Analysis of Randomized Controlled Trials," *Journal of Anxiety Disorders*, 61 (2019), 27–36.

347   Guang-Xin Wang and L. I. Li, "Virtual Reality Exposure Therapy of Anxiety Disorders," *Advances in Psychological Science*, 20 (2012), 1277–86.

348   Larissa Hjorth, Heather Horst, Sarah Pink, Baohua Zhou, Fumitoshi Kato, Genevieve Bell, Kana Ohashi, Chris Marmo, and Miao Xiao, "Digital Kinships: Intergenerational Locative Media in Tokyo, Shanghai and Melbourne," in *Routledge Handbook of New Media in Asia*, ed. Larissa Hjorth and Olivia Khoo (New York: Taylor and Francis, 2016), 251–62.

349   Margaret Gibson and Clarissa Carden, *Living and Dying in a Virtual World: Digital Kinships, Nostalgia, and Mourning in Second Life* (New York: Springer, 2018).

350   Richard L. Gilbert, Nora A. Murphy, and M. Clementina Ávalos, "Realism, Idealization, and Potential Negative Impact of 3d Virtual Relationships," *Computers in Human Behavior*, 27 (2011), 2039–46.

351   Debra Quackenbush, Jon G. Allen, and J. Christopher Fowler, "Comparison of Attachments in Real-World and Virtual-World Relationships," *Psychiatry*, 78 (2015), 317–27.

352   Yael Rosi Chen, Gurit E. Birnbaum, Jonathan Giron, and Doron Friedman, "Individuals in a Romantic Relationship Express Guilt and Devaluate Attractive Alternatives after Flirting with a Virtual Bartender," in *Proceedings of the 19th ACM International Conference on Intelligent Virtual Agents* (2019), 62–64.

353   Edward Schiappa, Mike Allen, and Peter B. Gregg, "Parasocial Relationships and Television: A Meta-Analysis of the Effects," *Mass Media Effects Research: Advances through Meta-Analysis* (2007), 301–14; Young Min Baek, Young Bae, and Hyunmi Jang, "Social and Parasocial Relationships on Social Network Sites and their Differential Relationships with Users' Psychological Well-Being," *Cyberpsychology, Behavior, and Social Networking*, 16 (2013), 512–17.

354   Mark Coulson, Jane Barnett, Christopher J. Ferguson, and Rebecca L. Gould, "Real Feelings for Virtual People: Emotional Attachments and Interpersonal Attraction in Video Games," *Psychology of Popular Media Culture*, 1 (2012), 176.

355   David Levy, "Falling in Love with a Companion," in *Close Engagements with Artificial Companions: Key Social, Psychological, Ethical, and Design Issues*, ed. Yorick Wilks (Amsterdam: John Benjamins, 2010).

356 Laura Lawton, "Taken by the Tamagotchi: How a Toy Changed the Perspective on Mobile Technology," *The iJournal: Graduate Student Journal of the Faculty of Information*, 2 (2017), 1–8.

357 Nicola S. Schutte and Emma J. Stilinović, "Facilitating Empathy through Virtual Reality," *Motivation and Emotion*, 41 (2017), 708–12.

358 Nahal Norouzi, Gerd Bruder, Jeremy Bailenson, and Greg Welch, "Investigating Augmented Reality Animals as Companions," in *2019 IEEE International Symposium on Mixed and Augmented Reality Adjunct (ISMAR-Adjunct)* (Beijing: IEEE, 2019), 400–403.

359 Thomas Chesney and Shaun Lawson, "The Illusion of Love: Does a Virtual Pet Provide the Same Companionship as a Real One?," *Interaction Studies*, 8 (2007), 337–42.

360 Cliffen Allen, Jeanny Pragantha, and Darius Andana Haris, "3d Virtual Pet Game 'Moar' with Augmented Reality to Simulate Pet Raising Scenario on Mobile Device," in *2014 International Conference on Advanced Computer Science and Information System* (IEEE, 2014), 414–19.

361 VidMid, "Latest in Vr Gives Users 'Girlfriend Experience Better Than the Real Thing,'" *VidMid* (2020), https://vidmid.com/news/latest-in-vr-gives-users-girlfriend-experience-better-than-the-real-thing?uid=177587.

362 Arne Dekker, Frederike Wenzlaff, Sarah V. Biedermann, Peer Briken, and Johannes Fuss, "VR Porn as 'Empathy Machine'? Perception of Self and Others in Virtual Reality Pornography," *The Journal of Sex Research*, 58 (2020), 1–6.

363 David Lafortune, Laurence Dion, and Patrice Renaud, "Virtual Reality and Sex Therapy: Future Directions for Clinical Research," *Journal of Sex & Marital Therapy*, 46 (2020), 1–17.

364 Amanda Sharkey and Noel Sharkey, "Granny and the Robots: Ethical Issues in Robot Care for the Elderly," *Ethics and Information Technology*, 14 (2012), 27–40; Kathleen Richardson, *An Anthropology of Robots and AI: Annihilation Anxiety and Machines* (New York: Routledge, 2015).

365 Noirin Curran, "Stereotypes and Individual Differences in Role-Playing Games," *The International Journal of Role-Playing*, 2 (2011), 44–58.

Chapter 8

366 Haosheng Ye, Hong Zeng, and Wendeng Yang, "Enactive Cognition: Theoretical Rationale and Practical Approach," *Acta Psychologica Sinica*, 51 (2019), 1270–80.

367 Matej Hoffmann and Rolf Pfeifer, "Robots as Powerful Allies for the Study of Embodied Cognition from the Bottom Up," *arXiv preprint arXiv:1801.04819* (2018).

368 George Lakoff and Mark Johnson, *Metaphors We Live By* (Chicago and London: University of Chicago Press, 2008).

369 Julia Prendergast, "Ideasthetic Imagining—Patterns and Deviations in Affective Immersion," *New Writing*, 18 (2020), 1–19.

370 Leopoldina Fortunati, "Robotization and the Domestic Sphere," *New Media & Society*, 20 (2018), 2673–90.

371 Walid Merrad, Alexis Héloir, Christophe Kolski, and Antonio Krüger, "Rfid-Based Tangible and Touch Tabletop for Dual Reality in Crisis Management Context," *Journal on Multimodal User Interfaces* (2021), 1–23.

372 Joel Chestnutt, Manfred Lau, German Cheung, James Kuffner, Jessica Hodgins, and Takeo Kanade, "Footstep Planning for the Honda Asimo Humanoid," in *Proceedings of the 2005 IEEE International Conference on Robotics and Automation* (IEEE, 2005), 629–34; Yoshiaki Sakagami, Ryujin Watanabe, Chiaki Aoyama, Shinichi Matsunaga, Nobuo Higaki, and Kikuo Fujimura, "The Intelligent Asimo: System Overview and Integration," in *IEEE/RSJ International Conference on Intelligent Robots and Systems* (IEEE, 2002), 2478–83.

373 Toru Ishida, "Development of a Small Biped Entertainment Robot Qrio," in *Micro-Nanomechatronics and Human Science, 2004 and The Fourth Symposium Micro-Nanomechatronics for Information-Based Society, 2004* (IEEE, 2004), 23–28.

374 C. A. Monje, Paolo Pierro, T. Ramos, Miguel González-Fierro, and Carlos Balaguer, "Modeling and Simulation of the Humanoid Robot Hoap-3 in the Openhrp3 Platform," *Cybernetics and Systems*, 44 (2013), 663–80.

375 Giorgio Metta, "Physical and Social Human-Robot Interaction," in *Proceedings of the 28th ACM Conference on User Modeling, Adaptation and Personalization* (2020), 3–3.

376 Kristiina Jokinen and Graham Wilcock, "Multimodal Open-Domain Conversations with the Nao Robot," in *Natural Interaction with Robots, Knowbots and Smartphones* (New York: Springer, 2014), 213–24.

377 IFR, "Service Robots Record: Sales Worldwide up 32%" (Frankfurt: The International Federation of Robotics, 2020).

378 Auxane Boch, Laura Lucaj, and Caitlin Corrigan, "A Robotic New Hope: Opportunities, Challenges, and Ethical Considerations of Social Robots" (Munich: Institute for Ethics in Artificial Intelligence, 2021).

379 Masahiro Mori, Karl F. MacDorman, and Norri Kageki, "The Uncanny Valley," *IEEE Robotics & Automation Magazine*, 19 (2012), 98–100.

380 Maike Paetzel-Prüsmann, Giulia Perugia, and Ginevra Castellano, "The Influence of Robot Personality on the Development of Uncanny Feelings," *Computers in Human Behavior*, 120 (2021), 106756; William D. Weisman and Jorge F. Peña, "Face the Uncanny: The Effects of Doppelganger Talking Head Avatars on Affect-Based Trust toward Artificial Intelligence Technology Are Mediated by Uncanny Valley Perceptions," *Cyberpsychology, Behavior, and Social Networking*, 24 (2021), 182–87.

381 Checking this in several ways showed that the trend is still solid. Indeed, similar results stem from other questions I analyzed that are identical in the two surveys, such as whether one uses a robot at home or at work. The percentage of those using robots at home increases over the years. The survey added "and artificial intelligence" to the question in 2017. This addition can change the results in either direction. But the overall trend seems warranted.

382   Yuefang Zhou and Martin H. Fischer, *AI Love You* (New York: Springer, 2019).

383   Christian J. A. M. Willemse and Jan B. F. van Erp, "Social Touch in Human–Robot Interaction: Robot-Initiated Touches Can Induce Positive Responses without Extensive Prior Bonding," *International Journal of Social Robotics*, 11 (2018), 285–304.

384   John-John Cabibihan, Deepak Joshi, Yeshwin Mysore Srinivasa, Mark Aaron Chan, and Arrchana Muruganantham, "Illusory Sense of Human Touch from a Warm and Soft Artificial Hand," *IEEE Transactions on Neural Systems and Rehabilitation Engineering*, 23 (2014), 517–27.

385   Jiaqi Nie, Michelle Park, Angie Lorena Marin, and S. Shyam Sundar, "Can You Hold My Hand? Physical Warmth in Human-Robot Interaction," in *2012 7th ACM/IEEE International Conference on Human-Robot Interaction (HRI)* (IEEE, 2012), 201–202.

386   Karen M. Grewen, Bobbi J. Anderson, Susan S. Girdler, and Kathleen C. Light, "Warm Partner Contact Is Related to Lower Cardiovascular Reactivity," *Behavioral Medicine*, 29 (2003), 123–30; Barbara Denison, "Touch the Pain Away: New Research on Therapeutic Touch and Persons with Fibromyalgia Syndrome," *Holistic Nursing Practice*, 18 (2004), 142–50; Sheldon Cohen, Denise Janicki-Deverts, Ronald B. Turner, and William J. Doyle, "Does Hugging Provide Stress-Buffering Social Support? A Study of Susceptibility to Upper Respiratory Infection and Illness," *Psychological Science*, 26 (2015), 135–47.

387   Alexis E. Block and Katherine J. Kuchenbecker, "Softness, Warmth, and Responsiveness Improve Robot Hugs," *International Journal of Social Robotics*, 11 (2019), 49–64.

388   Gurit E. Birnbaum, Moran Mizrahi, Guy Hoffman, Harry T. Reis, Eli J. Finkel, and Omri Sass, "Machines as a Source of Consolation: Robot Responsiveness Increases Human Approach Behavior and Desire for Companionship," in *2016 11th ACM/IEEE International Conference on Human-Robot Interaction (HRI)* (IEEE, 2016), 165–72.

389   Tobias Otterbring, "Smile for a While: The Effect of Employee-Displayed Smiling on Customer Affect and Satisfaction," *Journal of Service Management*, 28 (2017), 284–304.

390   Amy J. C. Cuddy, Caroline A. Wilmuth, Andy J. Yap, and Dana R. Carney, "Preparatory Power Posing Affects Nonverbal Presence and Job Interview Performance," *Journal of Applied Psychology*, 100 (2015), 1286; Emma Elkjær, Mai B. Mikkelsen, Johannes Michalak, Douglas S. Mennin, and Mia S. O'Toole, "Expansive and Contractive Postures and Movement: A Systematic Review and Meta-Analysis of the Effect of Motor Displays on Affective and Behavioral Responses," *Perspectives on Psychological Science* (2020), 1745691620919358.

391   Paul Ekman and Wallace V. Friesen, "The Repertoire of Nonverbal Behavior: Categories, Origins, Usage, and Coding," *Semiotica*, 1 (1969), 49–98.

392   Tass, "Russian Company Launches Serial Production of Androids" (Perm: Tass, Russian News Agency 2019).

393   Jinseok Woo, Janos Botzheim, and Naoyuki Kubota, "Facial and Gestural Expression Generation for Robot Partners," in *2014 International Symposium on Micro-Nanomechatronics and Human Science (MHS)* (IEEE, 2014), 1–6.

394   Alexis Stoven-Dubois, Janos Botzheim, and Naoyuki Kubota, "Fuzzy Gesture Expression Model for an Interactive and Safe Robot Partner," *arXiv preprint arXiv:1909.12054* (2019).

395   Zheng-Hua Tan, Nicolai Bæk Thomsen, Xiaodong Duan, Evgenios Vlachos, Sven Ewan Shepstone, Morten Højfeldt Rasmussen, and Jesper Lisby Højvang, "Isociobot: A Multimodal Interactive Social Robot," *International Journal of Social Robotics*, 10 (2018), 5–19.

396   Tino Lourens and Emilia Barakova, "User-Friendly Robot Environment for Creation of Social Scenarios," in *International Work-Conference on the Interplay between Natural and Artificial Computation* (New York: Springer, 2011), 212–21.

397   David Portugal, Paulo Alvito, Eleni Christodoulou, George Samaras, and Jorge Dias, "A Study on the Deployment of a Service Robot in an Elderly Care Center," *International Journal of Social Robotics*, 11 (2019), 317–41.

398   Serhan Cosar, Manuel Fernandez-Carmona, Roxana Agrigoroaie, Francois Ferland, Feng Zhao, Shigang Yue, Nicola Bellotto, and Adriana Tapus, "Enrichme: Perception and Interaction of an Assistive Robot for the Elderly at Home," *International Journal of Social Robotics*, 12 (2019), 779–805.

399   Berthold Bäuml, Florian Schmidt, Thomas Wimböck, Oliver Birbach, Alexander Dietrich, Matthias Fuchs, Werner Friedl, Udo Frese, Christoph Borst, and Markus Grebenstein, "Catching Flying Balls and Preparing Coffee: Humanoid Rollin'Justin Performs Dynamic and Sensitive Tasks," in *2011 IEEE International Conference on Robotics and Automation* (IEEE, 2011), 3443–44.

400   Will Knight, "Robot Dexterity," *MIT Technology Review*, 122 (2019), 11–12.

401   Celina Bertallee, "Global Study: 82% of People Believe Robots Can Support their Mental Health Better Than Humans" (Redwood Shores, CA: Oracle, 2020).

402   Byron Reeves and Clifford Ivar Nass, *The Media Equation: How People Treat Computers, Television, and New Media Like Real People and Places* (Cambridge, MA: Cambridge University Press, 1996).

403   Junchao Xu, Joost Broekens, Koen Hindriks, and Mark A. Neerincx, "Robot Mood Is Contagious: Effects of Robot Body Language in the Imitation Game," in *Proceedings of the 2014 International Conference on Autonomous Agents and Multi-Agent Systems* (Richmond, SC: International Foundation for Autonomous Agents and Multiagent Systems, 2014), 973–80.

404   Patricia E. Kahlbaugh, Amanda J. Sperandio, Ashley L. Carlson, and Jerry Hauselt, "Effects of Playing Wii on Well-Being in the Elderly: Physical Activity, Loneliness, and Mood," *Activities, Adaptation & Aging*, 35 (2011), 331–44.

405   Nancy Krieger, Kevin Smith, Deepa Naishadham, Cathy Hartman, and Elizabeth M. Barbeau, "Experiences of Discrimination: Validity and Reliability of a Self-Report Measure for Population Health Research on

Racism and Health," *Social Science & Medicine*, 61 (2005), 1576–96; María Ángeles Cea D'Ancona, "Measuring Xenophobia: Social Desirability and Survey Mode Effects," *Migration Studies*, 2 (2014), 255–80; Douglas P. Crowne and David Marlowe, "A New Scale of Social Desirability Independent of Psychopathology," *Journal of Consulting Psychology*, 24 (1960), 349; Anton J. Nederhof, "Methods of Coping with Social Desirability Bias: A Review," *European Journal of Social Psychology*, 15 (1985), 263–80.

406   Astrid M. Rosenthal-Von Der Pütten, Frank P. Schulte, Sabrina C. Eimler, Sabrina Sobieraj, Laura Hoffmann, Stefan Maderwald, Matthias Brand, and Nicole C. Krämer, "Investigations on Empathy Towards Humans and Robots Using Fmri," *Computers in Human Behavior*, 33 (2014), 201–12.

407   Isabelle M. Menne and Frank Schwab, "Faces of Emotion: Investigating Emotional Facial Expressions towards a Robot," *International Journal of Social Robotics*, 10 (2018), 199–209.

408   Alan D. A. Mattiassi, Mauro Sarrica, Filippo Cavallo, and Leopoldina Fortunati, "What Do Humans Feel with Mistreated Humans, Animals, Robots, and Objects? Exploring the Role of Cognitive Empathy," *Motivation and Emotion* (2021), 1–13.

409   Berat A. Erol, Abhijit Majumdar, Patrick Benavidez, Paul Rad, Kim-Kwang Raymond Choo, and Mo Jamshidi, "Toward Artificial Emotional Intelligence for Cooperative Social Human-Machine Interaction," *IEEE Transactions on Computational Social Systems*, 7 (2019), 234–46.

410   Isabel Pedersen, Samantha Reid, and Kristen Aspevig, "Developing Social Robots for Aging Populations: A Literature Review of Recent Academic Sources," *Sociology Compass*, 12 (2018), e12585; Sandra Bedaf, Gert Jan Gelderblom, and Luc De Witte, "Overview and Categorization of Robots Supporting Independent Living of Elderly People: What Activities Do They Support and How Far Have They Developed," *Assistive Technology*, 27 (2015), 88–100; Mariangela Manti, Andrea Pratesi, Egidio Falotico, Matteo Cianchetti, and Cecilia Laschi, "Soft Assistive Robot for Personal Care of Elderly People," in *2016 6th IEEE International Conference on Biomedical Robotics and Biomechatronics (BioRob)* (IEEE, 2016), 833–38; Denise Hebesberger, Tobias Koertner, Christoph Gisinger, and Jürgen Pripfl, "A Long-Term Autonomous Robot at a Care Hospital: A Mixed Methods Study on Social Acceptance and Experiences of Staff and Older Adults," *International Journal of Social Robotics*, 9 (2017), 417–29.

411   Chantal Cox-George and Susan Bewley, "I, Sex Robot: The Health Implications of the Sex Robot Industry" (London: British Medical Journal Publishing Group, 2018).

412   Chantal Cox-George and Susan Bewley, "Authors' Response to Dr John Eggleton's Letter," *BMJ Sexual & Reproductive Health*, 45 (2019), 169; John Eggleton, "Comment on 'I, Sex Robot: The Health Implications of the Sex Robot Industry,'" *BMJ Sexual & Reproductive Health* (2018).

413   Andrea Morris, "Meet the Man Who Test Drives Sex Robots," *Forbes* (September 27, 2018).

414 Christiane Eichenberg, Marwa Khamis, and Lisa Hübner, "The Attitudes of Therapists and Physicians on the Use of Sex Robots in Sexual Therapy: Online Survey and Interview Study," *Journal of Medical Internet Research*, 21 (2019), e13853.

415 Lynne Hall, "Sex with Robots for Love Free Encounters," in *International Conference on Love and Sex with Robots* (New York: Springer, 2016), 128–36.

416 Kristian Bankov, "Cultural Transformations of Love and Sex in the Digital Age," *Digital Age in Semiotics & Communication*, 2 (2019), 7–17.

417 David Levy, *Love and Sex with Robots: The Evolution of Human-Robot Relationships* (New York: Harper Collins, 2007).

418 Ian Pearson, "The Future of Sex: The Rise of the Robosexuals" (Essex: UK: Futurizon, 2015).

419 Sophie Wennerscheid, "Posthuman Desire in Robotics and Science Fiction," in *International Conference on Love and Sex with Robots* (New York: Springer, 2017), 37–50.

420 Robin Mackenzie, "Sexbots: Customizing Them to Suit Us Versus an Ethical Duty to Created Sentient Beings to Minimize Suffering," *Robotics*, 7 (2018), 70.

421 Lijun Zheng, Trevor A. Hart, and Yong Zheng, "The Relationship between Intercourse Preference Positions and Personality Traits among Gay Men in China," *Archives of Sexual Behavior*, 41 (2012), 683–89; Richard Lippa, "Gender-Related Individual Differences and the Structure of Vocational Interests: The Importance of the People–Things Dimension," *Journal of Personality and Social Psychology*, 74 (1998), 996; Richard A. Lippa, "Sex Differences in Personality Traits and Gender-Related Occupational Preferences across 53 Nations: Testing Evolutionary and Social-Environmental Theories," *Archives of Sexual Behavior*, 39 (2010), 619–36.

422 Nicola Döring and Sandra Poeschl, "Love and Sex with Robots: A Content Analysis of Media Representations," *International Journal of Social Robotics*, 11 (2019), 665–77.

423 Sally Whelan, Kathy Murphy, Eva Barrett, Cheryl Krusche, Adam Santorelli, and Dympna Casey, "Factors Affecting the Acceptability of Social Robots by Older Adults Including People with Dementia or Cognitive Impairment: A Literature Review," *International Journal of Social Robotics*, 10 (2018), 643–68.

424 Rebecca Q. Stafford, Bruce A. MacDonald, Xingyan Li, and Elizabeth Broadbent, "Older People's Prior Robot Attitudes Influence Evaluations of a Conversational Robot," *International Journal of Social Robotics*, 6 (2014), 281–97.

425 Laura Bishop, Anouk van Maris, Sanja Dogramadzi, and Nancy Zook, "Social Robots: The Influence of Human and Robot Characteristics on Acceptance," *Paladyn, Journal of Behavioral Robotics*, 10 (2019), 346–58.

426 Marcel Heerink, "Exploring the Influence of Age, Gender, Education and Computer Experience on Robot Acceptance by Older Adults," in *2011 6th ACM/IEEE International Conference on Human-Robot Interaction (HRI)* (IEEE, 2011), 147–48.

427   Elizabeth Broadbent, Rebecca Stafford, and Bruce MacDonald, "Acceptance of Healthcare Robots for the Older Population: Review and Future Directions," *International Journal of Social Robotics*, 1 (2009), 319.

428   Tatsuya Nomura, Takayuki Kanda, Tomohiro Suzuki, and Kensuke Kato, "Age Differences and Images of Robots: Social Survey in Japan," *Interaction Studies*, 10 (2009), 374–91.

429   Markus Blut, Cheng Wang, Nancy V. Wünderlich, and Christian Brock, "Understanding Anthropomorphism in Service Provision: A Meta-Analysis of Physical Robots, Chatbots, and Other AI," *Journal of the Academy of Marketing Science*, 49 (2021), 632–58.

Chapter 9

430   Yochanan E. Bigman, Adam Waytz, Ron Alterovitz, and Kurt Gray, "Holding Robots Responsible: The Elements of Machine Morality," *Trends in Cognitive Sciences*, 23 (2019), 365–68.

431   Andrea Owe and Seth D. Baum, "Moral Consideration of Nonhumans in the Ethics of Artificial Intelligence," *AI and Ethics* (2021), 1–12.

432   Cognilytica, *Worldwide AI Laws and Regulations 2020* (Washington, DC and Baltimore: Cognilytica, 2020).

433   Cara Tannenbaum, Robert P. Ellis, Friederike Eyssel, James Zou, and Londa Schiebinger, "Sex and Gender Analysis Improves Science and Engineering," *Nature*, 575 (2019), 137–46.

434   Rachel Adams and Nóra Ní Loideáin, "Addressing Indirect Discrimination and Gender Stereotypes in AI Virtual Personal Assistants: The Role of International Human Rights Law," *Cambridge International Law Journal*, 8 (2019), 241–57; Heather Suzanne Woods, "Asking More of Siri and Alexa: Feminine Persona in Service of Surveillance Capitalism," *Critical Studies in Media Communication*, 35 (2018), 334–49.

435   Steven J. Stroessner and Jonathan Benitez, "The Social Perception of Humanoid and Non-Humanoid Robots: Effects of Gendered and Machinelike Features," *International Journal of Social Robotics*, 11 (2019), 305–15.

436   Ryan Blake Jackson, Tom Williams, and Nicole Smith, "Exploring the Role of Gender in Perceptions of Robotic Noncompliance," in *Proceedings of the 2020 ACM/IEEE International Conference on Human-Robot Interaction* (2020), 559–67.

437   Youngme Moon Clifford Nass and Nancy Green, "Are Machines Gender Neutral? Gender-Stereotypic Responses to Computers with Voices," *Journal of Applied Social Psychology*, 27 (1997), 864–76.

438   Simone Alesich and Michael Rigby, "Gendered Robots: Implications for Our Humanoid Future," *IEEE Technology and Society Magazine*, 36 (2017), 50–59.

439   Laura Dattaro, "Bot Looks Like a Lady," *Slate* (2015), https://slate.com/technology/2015/02/robot-gender-is-it-bad-for-human-women.html.

440 Nicolaus A. Radford, Philip Strawser, Kimberly Hambuchen, Joshua S. Mehling, William K. Verdeyen, A. Stuart Donnan, James Holley, Jairo Sanchez, Vienny Nguyen, and Lyndon Bridgwater, "Valkyrie: NASA's First Bipedal Humanoid Robot," *Journal of Field Robotics*, 32 (2015), 397–419.

441 Sebastian Schneider and Franz Kummert, "Comparing Robot and Human Guided Personalization: Adaptive Exercise Robots Are Perceived as More Competent and Trustworthy," *International Journal of Social Robotics* (2020), 1–17.

442 Anna Pillinger, "Literature Review: Gender and Robotics" 13 (2019), 169–85; http://www.geecco-project.eu/fileadmin/t/geecco/Literatur/neu/Neu_30062020/Literatur_Review_Gender_and_Robotics.pdf.

443 Christoph Bartneck, Kumar Yogeeswaran, Qi Min Ser, Graeme Woodward, Robert Sparrow, Siheng Wang, and Friederike Eyssel, "Robots and Racism," in *Proceedings of the 2018 ACM/IEEE International Conference on Human-Robot Interaction* (2018), 196–204.

444 Robert Sparrow, "Robotics Has a Race Problem," *Science, Technology, & Human Values*, 45 (2020), 538–60.

445 Christoph Lutz, Maren Schöttler, and Christian Pieter Hoffmann, "The Privacy Implications of Social Robots: Scoping Review and Expert Interviews," *Mobile Media & Communication*, 7 (2019), 412–34.

446 Noriaki Mitsunaga, Christian Smith, Takayuki Kanda, Hiroshi Ishiguro, and Norihiro Hagita, "Adapting Robot Behavior for Human—Robot Interaction," *IEEE Transactions on Robotics*, 24 (2008), 911–16.

447 Francisco Erivaldo Fernandes, Guanci Yang, Ha Manh Do, and Weihua Sheng, "Detection of Privacy-Sensitive Situations for Social Robots in Smart Homes," in *2016 IEEE International Conference on Automation Science and Engineering (CASE)* (IEEE, 2016), 727–32.

448 Franziska Roesner, Tadayoshi Kohno, and David Molnar, "Security and Privacy for Augmented Reality Systems," *Commun. ACM*, 57 (2014), 88–96.

449 Jide S. Edu, Jose M. Such, and Guillermo Suarez-Tangil, "Smart Home Personal Assistants: A Security and Privacy Review," *arXiv preprint arXiv:1903.05593* (2019).

450 Hongbing Cheng and Gloria Rumbidzai Regedzai, "A Survey on Botnet Attacks," *American Scientific Research Journal for Engineering, Technology, and Sciences (ASRJETS)*, 77 (2021), 76–89.

451 Elaine Sedenberg, John Chuang, and Deirdre Mulligan, "Designing Commercial Therapeutic Robots for Privacy Preserving Systems and Ethical Research Practices within the Home," *International Journal of Social Robotics*, 8 (2016), 575–87.

452 Ceenu George, Mohamed Khamis, Emanuel von Zezschwitz, Marinus Burger, Henri Schmidt, Florian Alt, and Heinrich Hussmann, "Seamless and Secure VR: Adapting and Evaluating Established Authentication Systems for Virtual Reality," in *Network and Distributed System Security Symposium* (San Diego: NDSS, 2017).

453 Devon Adams, Alseny Bah, Catherine Barwulor, Nureli Musaby, Kadeem Pitkin, and Elissa M. Redmiles, "Ethics Emerging: The Story of Privacy and Security Perceptions in Virtual Reality," in *Fourteenth Symposium on Usable Privacy and Security ({SOUPS} 2018)* (2018), 427–42.

454 Martha C. Nussbaum, "Objectification," *Philosophy & Public Affairs*, 24 (1995), 249–91.

455 Piercosma Bisconti, "Will Sexual Robots Modify Human Relationships? A Psychological Approach to Reframe the Symbolic Argument," *Advanced Robotics*, 35 (2021), 1–11.

456 Joanna K. Malinowska, "What Does It Mean to Empathise with a Robot?," *Minds and Machines* 31 (2021), 361–76.

457 Papadaki, Evangelia (Lina), "Feminist Perspectives on Objectification," *The Stanford Encyclopedia of Philosophy* (Spring 2021 Edition), Edward N. Zalta (ed.).

458 Jaak Panksepp and Jules B. Panksepp, "Toward a Cross-Species Understanding of Empathy," *Trends in Neurosciences*, 36 (2013), 489–96.

459 Marta Osypinska, Michał Skibniewski, and Piotr Osypinski, "Ancient Pets. The Health, Diet and Diversity of Cats, Dogs and Monkeys from the Red Sea Port of Berenice (Egypt) in the 1st–2nd Centuries AD," *World Archaeology* 52:4 (2021), 639–53.

460 Brian Harrison, "Animals and the State in Nineteenth-Century England," *The English Historical Review*, 88 (1973), 786–820.

461 Piia Jallinoja, Markus Valtteri Vinnari, and Mari Niva, "Veganism and Plant-Based Eating as Political Consumerism," in *Oxford Handbook of Political Consumerism*, ed. Magnus Boström, Michele Micheletti, and Peter Oosterveer (Oxford: Oxford University Press, 2020).

462 Oliver Korn, Neziha Akalin, and Ruben Gouveia, "Understanding Cultural Preferences for Social Robots: A Study in German and Arab Communities," *ACM Transactions on Human-Robot Interaction (THRI)*, 10 (2021), 1–19.

463 Jennifer Robertson, *Robo Sapiens Japanicus: Robots, Gender, Family, and the Japanese Nation* (Los Angeles: University of California Press, 2017).

464 David J. Gunkel, *Robot Rights* (Cambridge, MA: MIT Press, 2018).

Conclusion

465 Kara C. Hoover, "Smell with Inspiration: The Evolutionary Significance of Olfaction," *American Journal of Physical Anthropology*, 143 (2010), 63–74.

466 James A. Coan, "Toward a Neuroscience of Attachment," in *Handbook of Attachment: Theory, Research, and Clinical Applications*, ed. Jude Cassidy and Philip R. Shaver (New York: The Guilford Press, 2008), 241–65.

467 Yuval Noah Harari, "Yuval Noah Harari: The World after Coronavirus," *Financial Times*, March 20, 2020.

Index

For the benefit of digital users, indexed terms that span two pages (e.g., 52–53) may, on occasion, appear on only one of those pages.

Note: Page references followed by an "f" indicate figure.

Aché, 30–31
active VR, 153
actroid, 13–14, 239
Advanced Telecommunications Research Institute International (Japan), 13–14
African apes, 37–38
age
  of artificial beings, 226–27
  elderly, robots and, 188–89, 195, 196
  at first marriage, rising, 86f, 88, 89, 90, 91
  human-AI relationships, acceptance of, and, 147–49
  robots, acceptance of, and, 216–18
agricultural society (Society 2.0), 5–6, 43.
    See also multigenerational family
  city in, 45–46
  domestication, animal, in, 46
  ethics in, 219–20
  food in, 45
  hierarchy in, 47–48, 51
  housing in, 44
  inheritance in, 51, 54–55
  labor in, 50–51
  law in, 50–51
  marriage in, 52–55
  money in, 49
  multigenerational family in, 6, 43–44, 51, 52–55, 96
  multigenerational family in, technology and, 56–57
  population growth in, 46
  property ownership, wealth accumulation in, 43–44, 46–48, 49, 50–55
  slavery in, 50–51
  technology in, development of, 44–46
  technology in, impact of, 46–51
  war in, 47
  weapons in, 45–46, 47

wheel in, 49
writing in, 48–49
AI. See artificial intelligence
aircraft, development of, 95–96
Alexa, 115–16
Alexander the Great, 49
algorithmic systems, flexibility of, 4
AlphaGo, 98
AlphaGo Zero, 98
Altered Carbon (Morgan, R.), 178, 179
Ancient Society (Morgan, L.), 39
Anki, 203–4
An Anthropology of Robots and AI (Richardson), 205
apes, African, 37–38
AR. See augmented reality
ARMAR (robot), 196–97
artificial beings
  age of, 226–27
  diversification of, 223–27
  education and ethics of, 227
  gender of, 223–25, 227
  objectification, human, and, 232–33
  privacy, security and, 228–30
  race of, 225–26, 227
  rights of, 235
artificial emotions, 108–12
artificial intelligence (AI), 16, 91. See also chatbots; human-AI relationships
  age in, 226–27
  apprehension about, 106–7, 128–29, 143–44
  attention mechanisms in, 119–20
  in cognitive revolution, 8–9, 97–99, 119–22
  dating study of, 139–40
  deep learning in, 117–18, 202
  development of, 116–19

artificial intelligence (AI) (cont.)
  disbelief in, suspending, 139
  embodied consciousness and, 178–80
  emotional intelligence of, 128–35
  emotional relationships with, 8–9, 113–
     16, 123, 127–35
  episodic memory in, 121
  ethics and, 221–22, 229–30, 236
  facial expressions understood by, 134–35
  familiarity with, VR compared to,
     159, 160f
  gender and, 214, 224
  general, strong, 97–98
  in Her, 105–6, 115, 123
  human relationships with, 114–15,
     135–43, 144, 145f, 146f, 157–58, 233
  imagination in, 121–22
  love and, 237
  mind-reading ability in, 141–43, 157–58
  multiple memory systems in, 120–21
  narrow, 97
  neural networks in, 97, 113, 116–17, 119–20
  PDP in, 117
  personal assistants, 8–9, 229–30, 236
  policy intervention for, 221
  privacy, security and, 228, 229–30
  progress in, 97–99
  in psychotherapy, 9, 129–34
  reinforcement learning in, 118–19, 198
  Relationships 5.0 and, 122–23
  Replika, 1, 113–15, 123, 127–28,
     144, 226–27
  robots using, 181, 202, 203–4, 228
  strategic games played by, 98–99, 121–22
  surveys, interviews on, 19, 20, 106–8, 122,
     123, 128–29, 142, 143–44, 221
  talking with, 123–28
  tolerance, support for, 107–8
  voice assistants, development of, 115–16
  Woebot, 9, 132–34
  XR compared with, 158
ASIMO (robot), 182–83
Asimov, Isaac, 182–83
Atlas (robot), 100
attention mechanisms, in AI, 119–20
augmented reality (AR), 153, 158–59
  MR and, 154
  sales of, rising, 10
  virtual pets through, 171–72
  VR and, 10, 99, 154

Avakin Life, 99–100, 171
avatars, virtual, relationships with, 168
Azar, 152

Babylon, ancient, 50, 53
Bacon, Francis, 59
Beer, Guy de, 125
biblum, 53–54
birth-control measures, 90
Bjorling, Elin, 218
Black Mirror (television series), 150–51, 173
BlenderBot, 125
BMI. See Brain-Machine Interface
BMJ. See British Medical Journal
body. See embodied cognition; embodied
     consciousness
bonobos, 31–32, 40
brain
  human-like, development of, 116–19
  in hunter-gatherer society, growth
     of, 26–27
Brain-Machine Interface (BMI), 150, 157–
     58, 229
brain-to-spine interface (BTSI), 157–58
Brick Dollbanger (customer
     reviewer), 207–8
British Medical Journal (BMJ), 206–7
BTSI. See brain-to-spine interface

Campaign Against Sex Robots, 205, 208
Cantillon, Richard, 65
Čapek, Karel, 182
Carden, Clarissa, 166–67
caring, by robots, 195–98
Catholic Church. See Church
CBT. See cognitive behavioral therapy
chatbots
  acceptance of, in survey, 144, 147
  BlenderBot code for, 125
  compassion for, 137–38
  conversational, 127–28
  developments in, 125
  disbelief in, suspending, 139
  Eliza, 129, 132
  human relationships with, 135–
     40, 142–43
  Insomnobot-3000, 126–27
  NLP in conversational, 127–28
  open-source codes for, 125
  psychotherapy, 129–30, 131–34

Replika, 1, 113–15, 123, 127–28, 144, 226–27
  Squirrel AI, 126
  task-oriented, 127
  Vincent, 137–38, 139, 142–43, 239
  Wysa, 131–32
chess, 98–99
chimpanzees, 31–32, 34–35
China, 67–68
Chou, Peter, 152
Church
  in industrial society, 61–62
  marriage and, 55
city, in agricultural society, 45–46
city, industrial. See industrial city
clan (Relationships 1.0), 6
  monogamy in, 36–37, 38, 39, 41
  non-monogamous relationships in, 33–34, 37–41, 42
  polyamory in, 39
  polyandry in, 33–34, 40
  polygamy, polyandry in, 33–34
  procreation in, 33–36, 42
  technology and, 41–42
Clark, Colin, 86, 87
clay, 44
cognitive behavioral therapy (CBT), 9, 129–30, 132
cognitive revolution, 8, 148–49, 239
  AI in, 8–9, 97–99, 119–22
  gender and, 214
  pathways to, 119–22
companions, robots as
  acceptance of, 20, 212f, 212–15, 213f, 216, 217–18
  romantic, sexual, 1, 13–14, 20, 97, 198–99, 205–12, 213f, 213–15, 216
companionship, marriage as, 55, 69
compartmentalization, of work, 66–67
compassion, 235–36
  for chatbots, 137–38
  cross-species, 233–34
  increase in, 233–35
  for robots, 235
  self-compassion, 137, 164–65
  VR and, 164–65
computer
  development of, 77
  in industry, 77–78
  personal, 77

serial computation, 116
  strategic games played by, 98
connection, safe. See safe connection
connectivity, 82–83
consciousness, embodied. See embodied cognition
conversational chatbots, 127–28
conversational skills, XR and, 175
COVID-19 pandemic, 5, 15, 151–52, 239
  mental health problems following, 16
  robots and, 12, 183, 204
  XR and, 152, 172
Crichton, Michael, 219
crossbreeding, 34–35
cross-species empathy, 233–34

Dactyl (robot), 198
Darwin, Charles, 59
data, 75–76, 79–80, 90
dating apps, 3–4
dating study, of AI, 139–40
David (robot), 100–1
Davis, Allison, 208–9
Deep Blue, 98
deepfakes, 119
deep learning, 117–18, 202
Deep Nostalgia, 119
deep Q-network (DQN), 121
Deep Reinforcement Learning, 118
Devol, George, 182
disbelief, suspending, 139
diversification, of artificial beings, 223–27
domestication
  animal, in agricultural society, 46
  human, in industrial society, 58
DQN. See deep Q-network
Durkheim, Émile, 71–72

EBT. See exposure-based therapy
education
  ethics and, 221, 227
  in information society, 89
Eggleton, John, 206–7
Egypt, 2011 protests in, 75
Egypt, ancient, 233–34
  marriage in, 54
  writing in, 48–49
elderly, robots and, 188–89, 195, 196
electromyography (EMG), 153
Eliza (chatbot), 129, 132

Ellie (virtual interviewer), 130
EMAR Project, 218
embodied cognition, 179
embodied consciousness
    AI and, 178–80
    ideasthesia and, 180
    XR and, 180
EMG. *See* electromyography
emotional attachment, XR and, 166–73
emotional intelligence
    of AI, 128–35
    of robots, 201–4
emotions, 238
    AI, emotional relationships with, 8–9,
        113–16, 123, 127–35
    artificial, 108–12
    in fictional content, by design, 108–11
    historical development of, 238
    interpreting, 134–35
    in love, 237, 238
    religious feelings, 111
    storytelling and, 165–66
    VR and, 163–67
    Woebot users on, 133–34
    XR and, 12, 163–73
empathy
    cross-species, 233–34
    increase in, 233–35
    individualism and, 83
    in information society, 83
    for robots, 198–201, 206, 235
    VR and, 165
Enlightenment, 61
Enlil (mythological figure), 53–54
ENRICHME project, 196
environmentalism, 234
episodic memory, 121
equality
    in hunter-gatherer society, 29–33, 47
    industrial society, social inequality
        in, 65–66
Erica (robot), 193–94, 239. *See also* actroid
ethics
    AI and, 221–22, 229–30, 236
    compassion, empathy, and, 233–34
    in connection, with artificial
        beings, 228–30
    cross-species, 233–34
    diversification, 223–27
    education and, 221, 227

environmental, 234
governments on emerging technologies
    and, 222
in history, 219–20
objectification and, 232–33
privacy, security, 228–31
Relationships 5.0 and, 220–21, 222–23,
    225–26, 227, 228–36
robots and, 219, 220–21, 223–26, 227,
    235, 236
Eugene Goostman (AI), 124
Europe, public opinion in
    Eurobarometer databases, 19, 147–48, 221
    on robots, 19, 185, 186*f*
exponential progress, 14–15
exposure-based therapy (EBT), 164
extended reality (XR)
    acceptance, apprehension of, 163, 174–77
    AI compared with, 158
    Avakin Life, 99–100, 171
    conversational skills and, 175
    Covid-19 pandemic and, 152, 172
    development, popularity of, 99–100, 155–62
    embodied consciousness and, 180
    emotions and, 12, 163–73
    gender and, 173, 174*f*, 175–76
    Google, 152, 158–59
    improvements in, 158, 159–62
    Microsoft, 154, 158–61
    objectification and, 173
    privacy, security and, 229–31
    in psychotherapy, 164–65, 169
    in *Ready Player One*, 155–58
    reality-based creations in, 154–55
    Relationships 5.0 and, 162–66
    religion and, 176
    rise of, 151–55
    in sensorial revolution, 10, 11, 12, 97, 99
    sexual relationships in, 172–73
    user-experience issues in, 159
    virtual characters, relationships with,
        162–63, 166–72

Facebook, 11–12, 78
    BlenderBot, 125
    Oculus, 151–52, 159–61, 171–72
    VR, AR wristband from, 153
facial expressions
    AI understanding, 134–35
    of robots, gestures and, 192–95

factories, 58, 60, 63
  conditions in, 66
  in industrial city, 58, 62–63, 66–68
  work compartmentalization and, 66–67
fertility treatments, 90
feudal society, 48, 51
fictional content, emotions in
  relationships and, 108–11
  religious feelings compared with, 111
fire, 26–27
Fisher, Allen, 86
food
  in agricultural society, 45
  in hunter-gatherer society, 26–27, 28–29
  in industrial society, 60, 63, 64
  robots preparing, 197–98, 212–13
Fourth Industrial Revolution, 7
From Sumer to Babylon (Glassner), 53

Game of Thrones (television series), 110
Garden of Eden, biblical story of, 43
Geminoid (robot), 13–14, 193
gender
  AI and, 214, 224
  of artificial beings, 223–25, 227
  cognitive revolution and, 214
  diversification and, 223–25
  objectification and, 173, 205
  robots, acceptance of, and, 212f, 212–14,
    213f, 216–17
  VR and, 173, 174f, 175–76
general (strong) AI, 97–98
Generative Pre-trained Transformer 2
  (GPT-2), 124–25
gestures
  human, studies of, 193
  of robots, facial expressions
    and, 192–95
Gibson, Margaret, 166–67
Gini Index, 29–30
Glassner, Jacques, 53
gloves, VR, 156–57
Go (game), 98
Google, 78
  Assistant, 115–16
  Cardboard, 152
  Go software, 98
  VR, 152
  XR, 152, 158–59
Go Touch VR haptic ring, 156–57

GPT-2. *See* Generative Pre-trained
  Transformer 2
Greece, ancient, 50–51

HAII. *See* Human-AI Interactions
Hames, Raymond, 40
Hammurabi, Code of, 50
Hand Arm System, 100–1
Hanson Robotics, 183
Happy Singlehood (Kislev), 85
HaptX glove, 156–57
Harmony (robot), 208, 209
healthcare
  chatbots in, 126–27
  robots in, 12–13, 16
Hebrew Scriptures, law in, 50
Henri II (French king), 55
Henry (robot), 208–9
Hepburn, Audrey, 183
Her (film), 105–6, 115, 123
hierarchy
  in agricultural society, 47–48, 51
  information society challenging, 84
Hill, Kim, 30–31
hippocampus, 121–22
HOAP (robot series), 183
HoloLens, 154, 159–61
Homo erectus, 26–27
Honda, 182–83
household, multigenerational family and, 56–57
housework, by robots, 196–98
housing
  in agricultural society, 44
  in hunter-gatherer society, 29, 33
  in industrial city, 67–69
HTC, 152
Huang, Jensen, 161–62
hugs, by robots, 191–92
Human-AI Interactions (HAII), 131
human-AI relationships, 136, 137
  acceptance of, 144–149, 145f, 146f
  age and acceptance of, 147–49
  chatbots, 135–40, 142–43
  dating study of, 139–40
  disbelief, suspending, in, 139
  mind-reading ability in, 141–43, 157–58
  Neuralink, 141, 142, 157–58
  Replika, romantic relationships with, 114–15
  studies on, 137–40
  as unhealthy, 136–37

human-like brain, development of, 116–19
humanoid robots. *See* robots
human-to-human relationships, 2, 17–18
hunter-gatherer society (Society 1.0), 5–6,
    25, 44. *See also* clan
  brain size growing in, 26–27
  clan in, 6, 33–42
  equality in, 29–33, 47
  fire in, 26–27
  food in, 26–27, 28–29
  housing in, 29, 33
  infanticide in, 36, 41
  language in, 27
  monogamy in, 36–37, 38, 39, 41
  non-monogamous relationships in, 33–
    34, 37–41, 42
  population control in, 35–36
  privacy absent in, 33
  procreation in, technology and, 33–36, 42
  in recent times, 30–31, 38, 39, 40
  relationships and technology in, 41–42
  relationships in, forms of, 39–41
  social organization of, 28–33, 42
Hyperconnect, 152

IBM, 124–25
iCub (robot), 183
ideasthesia, 180
IEEE. *See* Institute of Electrical and
    Electronics Engineers
imagination, in AI, 121–22
imagined communities, worlds, 10–11
individualism (Relationships 4.0)
  empathy and, 83
  in information society, 6, 76, 81–82, 83,
    84–91, 96, 104
  Internet and, 76, 81–82, 84, 88–89
  networked, 76, 81–82, 84, 88–89, 90, 96
  wireless phones and, 76, 83, 84
industrial city
  in China, contemporary, 67–68
  factories in, 58, 62–63, 66–68
  housing in, 67–69
  marriage and, 62–63, 64–65
  nuclear family in, 62–63, 64–65, 69–74, 96
  rise of, 62–64
  social fragmentation in, 64–69
  social inequality in, 65–66
  wealth accumulation and, 65
  work compartmentalized in, 66–67

industrial society (Society 3.0), 5–6. *See also*
    nuclear family
  domestication, human, in, 58
  ethics in, 219–20
  food in, 60, 63, 64
  Industrial Revolution beginning, 58
  knowledge in, 59–60
  marriage in, 62–63, 64–65, 69, 86*f*
  multigenerational family threatened
    by, 69–70
  nuclear family in, 6, 56–57, 58–59, 62–63,
    64–65, 69–74, 88, 96
  privacy in, 61–62
  religion in, 61–62
  secondary industry in, 86*f*, 87, 88
  space, time in, 60
  technology and, 59–60, 61–62
industry, computer in, 77–78
infanticide, 36, 41
information society (Society 4.0), 5–6. *See
    also* individualism
  data in, 75–76, 79–80, 90
  economy in, 79–81, 82–83, 86*f*, 87
  education in, 89
  empathy in, 83
  ethics in, 219–20
  hierarchy challenged in, 84
  individualism in, 6, 76, 81–82, 83, 84–91,
    96, 104
  knowledge in, 75–76, 84
  marriage in, 76, 85, 86*f*, 88
  as networked, 76, 81–82, 84, 88–89, 90, 96
  nuclear family, decline of, in, 76, 85–86,
    88–91, 96
  politics in, 81
  quaternary sector in, 86*f*, 87
  rise of, 77–81
  social changes in, 81–85
  solo living in, 85
  tertiary industry in, 86*f*, 87
inheritance
  in agricultural society, 51, 54–55
  legitimacy and, 54–55
Insomnobot-3000 (chatbot), 126–27
Institute of Electrical and Electronics
    Engineers (IEEE), 183–84, 185
Intel, 157
Internet
  connectivity, 82–83
  individualism and, 76, 81–82, 84, 88–89

rise of, 78, 81–82
singles and, 90
society and, 81–84, 90
intimacy, robots and, 205–12
iPhoneoid, 194–95
iRobot, robotic floor cleaner from, 196–97
Iroquois, 39
Ishiguro, Hiroshi, 13–14, 202–3
iSocioBot, 195

Japan
robots in, 13–14, 235
on Society 5.0, 6, 21
Jethá, Cacilda, 40
The Jetsons (television series), 187
Johnson, Mark, 179
Jonze, Spike, 105–6
Ju/'hoansi, 30
The Jungle (Sinclair), 233–34
Justin (robot). *See* Rollin' Justin

kami, 235
Kami Computing, 125, 206
Kim, Jinwoo, 131
Kislev, Elyakim, 85
knowledge
in industrial society, 59–60
in information society, 75–76, 84
Kurzweil, Ray, 14–15
Kuyda, Eugenia, 113–14

labor, in agricultural society, 50–51
Lakoff, George, 179
language, in hunter-gatherer society, 27
law, in agricultural society, 50–51
learning
in AI, 117–19, 198, 202
deep, 117–18, 202
reinforcement, 118–19, 198
by robots, 198, 202–4
legitimacy, inheritance and, 54–55
Levy, David, 206
Lindner, Philip, 162–63, 168–69
Living and Dying in a Virtual World (Gibson and Carden), 166–67
loneliness, 16
love, 2, 237, 239
AI, robots, and, 237
components of, 102–4
emotions in, 237, 238

historical development of, 238
neural systems and, 238
Love and Sex with Robots (Levy), 206
Luka, 1

Mackenzie, Robin, 211
Maia (chess engine), 98–99
marriage
age at first, rising, 86f, 88, 89, 90, 91
in agricultural society, 52–55
as companionship, emergence of, 55, 69
in industrial society, 62–63, 64–65, 69, 86f
in information society, 76, 85, 86f, 88
property ownership and, 52–55
Marx, Karl, 30–31
Mazurenko, Roman, 113
McArthur, Neil, 214, 225
The Media Equation (Reeves and Nass), 199
medieval period
feudal society of, 48, 51
labor in, 51
merchant code of, 50
mental health, 16
merchant code, medieval, 50
Mesopotamia, ancient, 48, 49, 50, 53–54
Metaphors We Live By (Lakoff and Johnson), 179
metaverses, 161–62
Microsoft, 124–25
HoloLens, 154, 159–61
XR, 154, 158–61
mind-reading ability, in AI
human-AI relationships and, 141–43, 157–58
Neuralink, 141, 142, 157–58
Vincent and scenario of, 142–43
mixed reality (MR), 154, 185
Mobileye, 115, 128
Moley Robotics, 197–98
Monet, Claude, 117–18
money, in agricultural society, 49
monogamy
in clan, 36–37, 38, 39
human inclination to, questioning, 37, 38–39
in hunter-gatherer society, 36–37, 38, 39, 41
parental care and, 38
social, 41
Moore, Gordon, 14–15

Moore's law, 14–15
morality. *See* ethics
Morgan, Lewis, 39
Morgan, Richard K., 178, 179
Moyle, Wendy, 188–89
MR. *See* mixed reality
Mubarak, Hosni, 75
multigenerational family (Relationships
    2.0), 6, 45
  agrarian technology and, 56–57
  household and, 56–57
  industrial society threatening, 69–70
  inheritance in, 51, 54–55
  marriage in, 52–55
  nuclear family replacing, 69–74, 96
  property ownership and, 43–44, 51–52, 56
multiple memory systems, AI, 120–21
Mumford, Lewis, 46
Musk, Elon, 141, 157–58, 222
My Heritage, 119

Nao (robot), 183
narrow AI, 97
NASA, 224–25
Nass, Clifford Ivar, 199
Natufians, 45
Natural Language Processing
    (NLP), 127–28
Neolithic period (New Stone Age), 43–
    44, 45, 46
Netflix, 110
networked, information society as. *See*
    information society
networked individual. *See* individualism
Neuralink, 141, 142, 157–58
neural networks, 97, 113, 116–17, 119–20
neural systems, love and, 238
New Stone Age. *See* Neolithic period
Nietzsche, Friedrich, 61
NLP. *See* Natural Language Processing
non-monogamous relationships
  in hunter-gatherer society, 33–34, 37–
    41, 42
  polyamory, 39, 40
  polyandry, 33–34, 37, 39, 40
  polygyny, 40–41
nuclear family (Relationships 3.0), 17
  in industrial city, 62–63, 64–65, 69–74, 96
  in industrial society, 6, 56–57, 58–59, 62–
    63, 64–65, 69–74, 88, 96

in information society, decline of, 76, 85–
    86, 88–91, 96
  marriage and, 70–71
  multigenerational family replaced by,
    69–74, 96
  privacy and, 62
  rise of, 56–57, 62, 70–74
nursing, by robots, 195–98
Nussbaum, Martha, 232

objectification
  artificial beings and human, 232–33
  ethics and, 232–33
  gender and, 173, 205
  Nussbaum on, 232
  robots and human, 232
  sex robots and, 205
  XR and, 173
Octavia (robot), 101
Oculus, 159–61, 171–72
  Oculus Quest, 151–52
  Oculus Quest 2, 151–52
Omni treadmills, 156
O'Neill, John, 63
OpenAI, 198
open-source codes, for chatbots, 125
On the Origin of Species (Darwin), 59

parallel distributed processing (PDP), 117
parasocial relationships, 169–70
parental care, monogamy and, 38
partible paternity, 40
passive VR, 153
PDP. *See* parallel distributed processing
Pearson, Ian, 210–11
Pega, AI survey by, 122, 143–44
personal assistants, AI, 8–9, 229–30, 236
personal computer, 77
personalization, 228–29
pets, virtual, 171–72
physical assistance, robots for, 212*f*, 212–13
physical closeness, robots and, 189–92
physical revolution, 8, 239
  acceptance of, opposition to, 147–48, 185,
    186*f*, 212*f*, 212–18, 213*f*
  gender and acceptance of, 212*f*, 212–14,
    213*f*, 216–17
  robots in, 12–14, 97, 99, 100–2, 181–85
physiological signs, emotions interpreted
    with, 135

Play, Frederic Le, 71
policy intervention, 221
politics, in information society, 81
polyamory, 39–40
polyandry, 33–34, 37, 39, 40
polygamy, 39
polygyny, 40–41
population
    in agricultural society, growth of, 46
    in hunter-gatherer society, control of, 35–36
post-traumatic stress disorder (PTSD),
        130–31, 164
primary industries, 86f, 86
primitive communism, 30–31
privacy, 231
    artificial beings and, 228–30
    ethics of security and, 228–31
    in hunter-gatherer society, absent, 33
    in industrial society, 61–62
    nuclear family and, 62
    VR and, 230–31
    XR and, 229–31
procreation, 33–36, 42
progress, exponential, 14–15
Project Debater, 124–25
Promobot, 194
property ownership
    in agricultural society, 43–44, 46–48,
        49, 50–55
    marriage and, 52–55
    multigenerational family and, 43–44,
        51–52, 56
    writing and, 48
psychotherapy
    AI in, 9, 129–34
    CBT, 9, 129–30, 132
    VR in, 164–65, 169
PTSD. See post-traumatic stress disorder

Q (genderless voice), 224
Qrio (robot), 183
quaternary sector, 86f, 87

race, of artificial beings, 225–26, 227
Radford, Nicolaus, 224–25
RDW. See redirected walking
Ready Player One (film)
    BMI in, 157–58
    retina-based technology in, 157
    VR in, 155–58

Realbotix, 205, 208–9
reality-based creations, in XR, 154–55
Reclaiming Conversation (Turkle), 175
redirected walking (RDW), 156
Reeves, Byron, 199
Reformation, 62
reinforcement learning, 118–19, 198
relationships, components of, 102–5
relationships, with virtual characters. See
    virtual reality
Relationships 1.0. See clan
Relationships 2.0. See
    multigenerational family
Relationships 3.0. See nuclear family
Relationships 4.0. See individualism
Relationships 5.0. See specific topics
religion
    in industrial society, relationships
        and, 61–62
    religious feelings, 111
    VR and, 176
Replika (chatbot), 1, 114, 115
    acceptance of, in survey, 144
    age and, 226–27
    development of, 113–14
    NLP and, 127–28
    romantic relationships with, 114–15
    talking with, users on, 123, 128
research methods, 18–21
retina-based technology, VR, 157
Richardson, Kathleen, 205–6, 208
ring-tailed lemurs, 34–35
Robertson, Jennifer, 235
Robo-C (robot), 194
Robonaut (robot), 224
robosexuality, 210–11
Robot Companions for Citizens, 211
RobotCub Consortium, 183
robotic floor cleaner, from iRobot, 196–97
robots, 91, 100. See also sex robots; social
    robotics
    acceptance of, as companions, 20, 212f,
        212–15, 213f, 216, 217–18
    acceptance of, opposition to, 147–48, 185,
        186f, 187–89, 199, 212–18
    actroid, 13–14, 239
    age and acceptance of, 216–18
    age of, 226–27
    AI used by, 181, 202, 203–4, 228
    caring and nursing by, 195–98

robots (*cont.*)
   Covid-19 pandemic and, 12, 183, 204
   developments in, 99, 100–2
   diversification of, 223–27
   elderly and, 188–89, 195, 196
   emotional intelligence of, 201–4
   empathy, compassion for, 198–201,
      206, 235
   Erica, 193–94, 239
   ethics and, 219, 220–21, 223–26, 227,
      235, 236
   European public opinion on, 19, 185, 186*f*
   facial expressions, gestures of, 192–95
   food prepared by, 197–98, 212–13
   Geminoid, 13–14, 193
   gender and acceptance of, 212*f*, 212–14,
      213*f*, 216–17
   gender of, 223–25
   Harmony, 208, 209
   in healthcare, 12–13, 16
   Henry, 208–9
   housework by, 196–98
   humanoid, development of, 182–84
   in Japan, 13–14, 235
   learning by, 198, 202–4
   love and, 237
   MR technology with, 185
   objectification and, 205, 232
   personalization of, 228–29
   for physical assistance, acceptance of,
      212*f*, 212–13
   in physical revolution, 12–14, 97, 99,
      100–2, 181–85
   privacy, security and, 228–30
   race of, 225–26
   Realbotix, 205, 208–9
   Relationships 5.0 and, 187–89,
      201, 211–12
   romantic relationships with, 1, 13–14, 20,
      97, 198–99, 211, 213*f*, 213–15, 216
   sales of, market for, 12, 183–84
   sex, debates on, 205–8, 209–10, 211–12
   sex, intimacy, and, 205–12
   sexuality of, 227
   social robotics, 181–85, 187–92, 195–
      98, 218
   Sophia, 183, 184
   as spouses, humanoid, 1, 210–11
   term for, 182
   touch, physical closeness of, 189–92

   uncanny valley problem of, 184, 185,
      188, 192
   uprising of, 220
   Valkyries, 224–25
   Vector, 203–4
   on *Westworld*, 219, 220
Rogerian psychotherapy, 129
Rollin' Justin (robot), 197
Romans, 48
romantic relationships
   with Replika, 114–15
   with robots, 1, 13–14, 20, 97, 198–99, 211,
      213*f*, 213–15, 216
   in VR, 150–51
Rossum's Universal Robots (Čapek), 182
Royal Society convention (2014), 124
Ruggles, Steven, 73–74
Ryan (chatbot), 126–27
Ryan, Christopher, 40

safe connection
   with artificial beings, 228–30
   ethics of, 228–31
   with XR, 229–31
Samantha (robot), 13
secondary industry, 86*f*, 86, 88
Second Life, 166–68
secret marriages, 55
security
   AI and, 228, 229–30
   artificial beings and, 228–30
   ethics of privacy and, 228–31
   XR and, 229–31
self-compassion, 137, 164–65
sensorial revolution, 8, 10, 239
   acceptance, apprehension of, 163, 174–77
   imagined communities, worlds, in, 10–11
   VR and AR in, 10, 99
   XR in, 10, 11, 12, 97, 99
serial computation, 116
service sector, 86*f*, 87
Sex at Dawn (Ryan, C., and Jethá), 40
SexLikeReal, 172
sex robots, 13, 97
   acceptance of, 213*f*, 213–15, 216
   customer reviews on, 207–8
   debates on, 205–8, 209–10, 211–12
   Realbotix, 205, 208–9
   robosexuality, 210–11
   sex therapists, physicians on, 209–10

sexuality, of robots, 227
sexual relationships, in XR, 172–73
Shakey (robot), 182
Shashua, Amnon, 115, 128
Shinto, 235
Silicon Coppélia, 139–40
Sinclair, Upton, 233–34
singles, Internet and, 90
Siri, 115–16
slavery, 50–51
smartphones, 83, 141–42
    introduction of, 18, 78–79
    iPhoneoid, 194–95
    privacy and, 229–30
social breakdown, theory of, 72
social fragmentation, 64–69
social inequality, 65–66
social monogamy, 41
social networking tools, 82, 152
social robotics
    acceptance of, 218
    caring and nursing in, 195–98
    Relationships 5.0 and, 187–89
    rise of, 181–85
    touch, physical closeness in, 189–92
Society 1.0. See hunter-gatherer society
Society 2.0. See agricultural society
Society 3.0. See industrial city; industrial
    society
Society 4.0. See information society
Society 5.0
    ethics and, 221
    Japanese government coining, 6, 21
    policy intervention for, 221
    relationships in, 6–7, 14–18
    as Super Smart, 6, 96–97
    technological breakthroughs in, speed
        of, 101–2
    term coined, 6
softness, of robots, 190
solo living, 85
Solo software, 154–55
Sophia (robot), 183, 184
South Korea, 235
space, in industrial society, 60
Sparrow, Robert, 226
Spielberg, Steven, 155
spouses, robots as, 1, 210–11
Squirrel AI (chatbot), 126
Starkweather, Katherine, 40

Star Wars films, 182
storytelling, 165–66
strategic games, 98–99, 121–22
Striking Vipers (fictional game), 150–
    51, 157–58
strong AI. See general AI
Sud (mythological figure), 53–54
Sumer, 48, 53
Super Smart society, 6, 96–97
Suzman, James, 30

Tamagotchi, 170–71
task-oriented chatbots, 127
technological breakthroughs, speed
    of, 101–2
technology, defining, 25
temperature, of robots, 190–91
tertiary industry, 86f, 87
Teslasuit, 157
TetaVi, 154–55
three-dimensional (3D) printing, 7–8
time, in industrial society, 60
Tinder, 4
Tom (AI), 139–40
touch, robots and, 189–92
treadmills, VR, 156
Trollope, Frances, 66
Turing, Alan, 116–17, 124
Turing test, 124–25, 206
Turkle, Sherry, 175

uncanny valley, 184, 185, 188, 192
Unimate (robot), 182
users
    for chatbots, compassion of, 137–38
    on Replika, talking with, 123, 128
    with Vincent, bonding, 138
    in Woebot trial, 132–34
    in XR, user-experience issues, 159

Valkyrie (robot), 224–25
Variable Stiffness Actuators (VSA), 100–1
Vaunt, 157
Vector (robot), 203–4
Verdicchio, Mario, 138
vibration technology, VR, 156–57
Vincent (chatbot), 142–43, 239
    compassion and, 137–38
    scenarios created by, 137–38, 139
    users bonding with, 138

virtual pets, 171–72
virtual reality (VR). *See also* extended reality
  acceptance of, 163
  active and passive, 153
  AR and, 10, 99, 154
  on *Black Mirror*, 150–51, 173
  BMI in, 150, 157–58
  companies developing, 11–12
  compassion, empathy, and, 164–65
  EBT with, 164
  elements of, 152–54
  emotions and, 163–67
  Facebook and, 11–12, 153
  familiarity with, AI compared to,
    159, 160*f*
  gender and, 173, 174*f*, 175–76
  improvements in, 159–61
  popularity of, 10, 91, 151–54
  privacy, security and, 230–31
  in psychotherapy, 164–65, 169
  RDW in, 156
  in *Ready Player One*, 155–58
  religion and, 176
  retina-based technology in, 157
  rise of, 151–54
  romantic relationships in, 150–51
  sales of, rising, 10, 91
  sexual relationships in, 172
  storytelling and, 165–66
  treadmills, 156
  user-experience issues in, 159
  vibration technology in, 156–57
  virtual characters, relationships with,
    162–63, 166–72
voice assistants, 115–16

VR. *See* virtual reality
VRChat, 171–72
VSA. *See* Variable Stiffness Actuators

war, 47
warmth, of robots, 190–91, 205
wealth accumulation
  in agricultural society, 43, 46, 47, 48
  industrial city and, 65
  writing and, 48
weapons, 45–46, 47
Westworld (television series), 219, 220
WhatsApp, 125
wheel, 49
wireless phones, 79
  individualism and, 76, 83, 84
  smartphones, 18, 78–79, 83, 141–42, 194–
    95, 229–30
Woebot, 9, 132
  emotional impact of, users on, 133–34
  trial of, 132–34
  user complaints about, 134
women, education of, 89
Work (Suzman), 30
Wright brothers, 95–96
writing, in agricultural society, 48–49
Wysa (chatbot), 131–32

X2 AI, 131
XR. *See* extended reality
XRSpace Manova, 152

Ying-Ying (robot), 1, 210–11

Zheng Jiajia, 1, 210–11